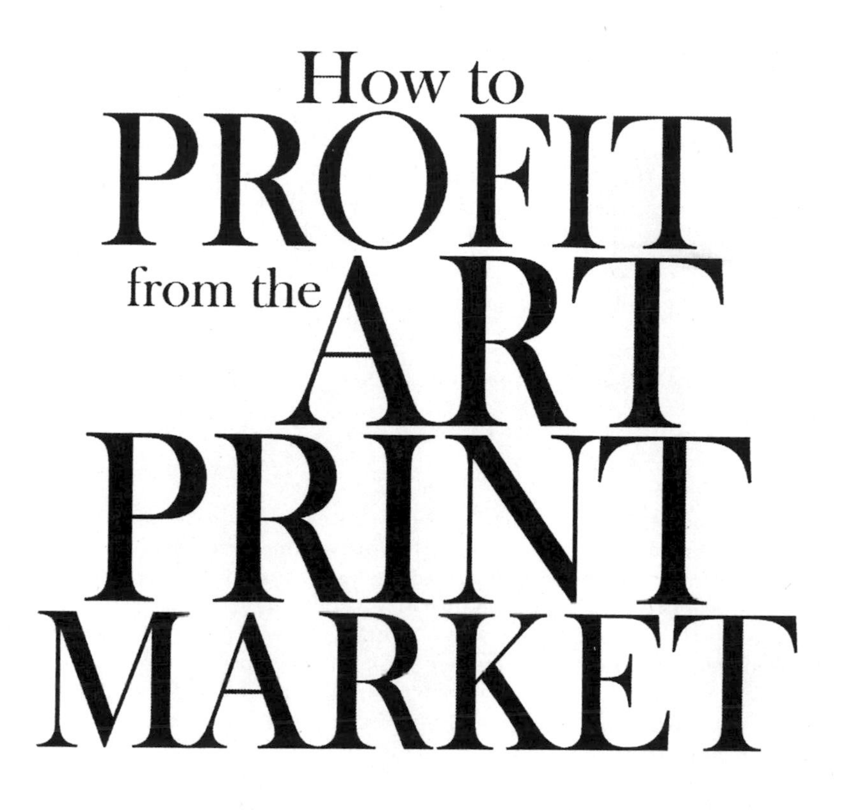

How to PROFIT from the ART PRINT MARKET

Creating Cash Flow From Original Art

Practical Advice for Visual Artists

Barney Davey

Legal Notices

DISCLAIMER
The information contained in this book is intended for general guidance on matters of interest only. Although the author has extensively researched the materials in this book to ensure accuracy, the author assumes no liability or responsibility for its use, reliance on its content, or reliance whatsoever. Nor does the author assume any liability or responsibility for errors, inaccuracies, omissions, misinterpretations, or any inconsistency contained, and specifically disclaims all possible liability premised upon reliance or use of this book. This book is not offered nor should readers rely upon it as legal, financial or other professional advice. In all legal, financial and other professional questions, it is prudent and incumbent upon the reader to retain an attorney and competent financial and other appropriate advisers to be consulted regarding specific questions as to how those questions relate to specific instances. Accordingly, the information in this book is provided with the understanding that the author is not herein engaged in rendering legal, financial or other professional advice and services. As such, it should not be used as a substitute for consultation with competent legal or professional advisers.

CREDITS: Bookcovers.com for cover & interior page design

ISBN 0-9769607-0-2
Library of Congress Control Number: 2005904324

Printed in U.S.A. by Lightning Source, Inc.

Published 2005

Table of Contents

Introduction

> From the moment I picked your book up until I laid it down, I was convulsed with laughter. Someday I intend reading it.
>
> – Groucho Marx

I promise if you will be unlike Groucho and grace me by reading this book, you will come away with a clearer understanding of the art print market and you'll be better equipped to make good decisions and prosper in it. From personal observations, I know artists in the art print market can earn great rewards and recognition. This book is written to help guide you to success as a print artist.

Gaining an understanding of the nature of the print business may be critical for a visual artist who wishes to enjoy success in his or her career. While there are numerous excellent books on the subject of the business of being an artist and on art marketing, virtually none of them provide more than a couple of paragraphs or pages, or in some cases nothing, regarding the ins and outs of the print market. As getting into print is the best way for visual artists to generate a secondary source of income from their creative efforts, this book is essential for those who want to learn how to profit from the art print market.

My 15-year tenure with *DECOR* magazine and its sister tradeshow, Decor Expo, gave me access and allowed me to observe many of the industry's most successful print publishers, both self-published and established catalog publishers. My intentions with this book are to distill those experiences and share my observations with visual artists.

Artists seeking to break into the print market would call me for advice on getting into the print business in the years I worked for *DECOR*. Since having repeat customers was my primary means of being successful, it was important for me to be able to provide new artists with useful, practical information to help them could make informed decisions about how to best proceed.

If the readers of this book gain a better grasp of the art print market and with it the ability to judge the best course of action for them based on their capabilities and resources, my aspirations for it will be met. Best wishes to all the intrepid souls who embark upon an art print market career. Send me a postcard from the top!

A grateful nod to my wife, Mary, for her love, her patience and support, and to Bill Vernor for his perceptive insights which improved the book.

Barney Davey

> To see far is one thing, going there is another.
>
> – Constantin Brancusi

Chapter One
Goals and Vision

> Every artist was first an amateur.
>
> – Ralph Waldo Emerson

The primary goal of this book is to help you gain an understanding of how to become a successful published artist whose original works are selling in some reproductive multiple form in the art print marketplace. As publishing is not for every artist, it can also serve as a sounding board to help you decide, or at least get closer to making an informed decision, whether or not publishing your art as a self-publisher, or in concert with an established art print publisher, is the path for you to take.

Defining the Term Print

Before we go farther, let's allow that the art world is one where confusing terms and contradictions exist. Since the word "print" is used in the title of this book, and it is a prime example of a term with confusing yet synonymous meanings in the art world, it is imperative to provide a clear definition of what a "print" is as it pertains to this book.

The terms "print or prints" in this book are meant to discuss those works of art on paper or canvas that primarily are reproductions of originals and which are always sold in multiples. Typically, these prints are reproductions of original art works including oils, pastels, acrylics and watercolors. An exception is some prints are not reproductions, but are created in their origin as prints. Examples are computer-generated images, monotypes and serigraphs designed to be prints.

I acknowledge there are many other methods of creating art on paper, but emphasize here that this book above all concerns itself with, but is not limited to, posters, open edition prints, canvas transfers, limited edition offset prints, giclées and serigraphs suited to be marketed to the mainstream populace.

Another use of the term "print" comes from a body of artists, educators, gallerists, museums and collectors whose interest revolves around what I will call the printmaking community. Printmaking is an ancient art and is still taught at universities where students major in printmaking.

Many in the printmaking community who specialize in media as linotypes, woodcuts, monotypes, serigraphs, etchings, collographs, aquatints, mezzotints and so forth, take a more elite connotation of the word print. They often eschew those prints made for the mainstream market. In this book I will use the term "elite prints" to convey those works of art on paper that come from the printmaking community.

It becomes obvious then one can hear the word print used to describe some piece of art and without further clarification be clueless as to what the word means or how the print was made. Are we describing an open edition lithographic offset reproduction printed yesterday, or an original graphic etching made from a plate created by Rembrandt 300 years ago? It's not wrong for the term print to be used to identify art on paper that are of different origins, but it is confusing without further explanation.

What Is A Dealer And How Are They Different From Retailers Or Gallerists?

You will see the word "dealer" used throughout this book. For our purposes and within the industry in general, dealers are those who buy art at wholesale and sell at retail. The term encompasses not just gallerists who would be brick & mortar art retailers, but also includes art consultants, high-end interior designers and others who may not have a retail location but who own, trade, buy and consign art, and who have the capability to exert great influence over buying decisions by major collectors, museums and corporate art buyers.

The contemporary art market, fairly or not, can be broken down from a price point of view into three tiers. Contemporary here is not used to describe a particular style of art, but rather to denote those pieces of art that are currently being sold and marketed in a high-to-low price continuum from:

1. High-end fine art galleries, museums, auction houses and important dealers and interior designers typically found in major metropolitan areas, resorts and at the high-end consumer art shows including Biennale Venice, Art Basel and the Palm Beach Art Show. Expect to see them selling fine art originals, elite prints and sculpture;

2. Middle market galleries, art consultants and dealers and some interior designers selling higher priced limited edition giclées, serigraphs, sculptures and offset limited edition prints and moderately priced originals;

3. Poster and frame shops, lower end galleries, mass market retailers, furniture stores and

contract design firms selling posters, open edition prints and some lower priced offset limited edition prints and giclées.

This book aims to help those who wish to see their works sold in the middle and lower tiers of the market, which if viewed as a pyramid would form the base and bulk of the art market in terms of numbers of buyers. There are potential tremendous rewards and recognition from serving these areas of demand for art. And, contrary to what many think there are plenty of examples of artists who have had it all, meaning they have enjoyed representation at chic Manhattan galleries and benefited from having their work reproduced and sold to the mass market.

Calvin Goodman, in his 531-page tome *Art Marketing Handbook* which was the bible for art marketers for decades, long maintained the print market, as it is considered in this book, was not a great opportunity for economic gain for artists.

Goodman went on to say he was proven wrong. He acknowledges that an artist he watched for two decades, Arthur Secunda, while not becoming wealthy, managed to nevertheless do well with poster publisher Haddad's Fine Arts. *www.haddadsfinearts.com*

According to Goodman, Secunda found being in posters helped him maintain his studios in France and California and he believed the exposure did not hurt his original print and collage market. In fact, Goodman reports that Secunda considered his posters not only spawned imitators, but also helped to popularize his name and imagery.

There are no guarantees that reading this book you will achieve the status of an Arthur Secunda, or simultaneously secure a lofty position in the top tier and the lower tier of the market as he did. You may be assured you will gain a greater understanding and appreciation of how the art print market operates in the middle and lower tiers. You will learn about the similarities in traits and attributes that artists already successful in this market share and will come away with a clearer decision on how you can best benefit from the potential the market offers.

As you'll come to see, the decision to publish is the first of many you must make along the way to earning reward and recognition from having your work sold in the print market. The decisions you make will put you on one of the many paths to print market success for artists.

I wish it were possible to provide you with all the ways artists have made nice careers in the print market. But this would not be a book, but an encyclopedia requiring far too much shelf space and time to read. Instead, I'll give you the best examples of some I have known or whose careers I have had the opportunity to study.

To start, we'll explore these questions and sentiments in this chapter:

- What do you want to achieve with your art?
- Do you have a mission statement?
- Is commercial success included in your vision for personal achievement?
- Ambition is not a dirty word
- Does the thought of high integrity, high sales and high bank account as an alternative to the starving artist syndrome appeal to you?
- Is a career that offers potential high repute but almost assured low bank balance worth pursuing?

For those of you who desire to put yourself on the road to publishing, this guidebook will help you decide which of the many forks in the road you should take that lead to seeing your art published. At this point, the temptation is too great to resist quoting an esteemed philosopher who said these things:

> When you come to a fork in the road, take it.
>
> You got to be careful if you don't know where you're going, because you might not get there.
>
> - Yogi Berra

My goal here is to be more accommodating than Yogi's advice in that I intend to help you make good decisions about which of the many forks you'll encounter upon entering the art print market and to help you get somewhere with your art print career.

It would be hard to imagine you took the time and spent the money to find and read this book if you didn't already have the idea of making some money, if not a full-fledged business from your art. You are encouraged to make pursuing your goals a journey and not a destination from whatever fork in the road you choose.

What Do You Want To Achieve With Your Art?

The choices begin with you defining what sort of artist you are or want to be regarding your success on a commercial level. If you plan to sell your work, you can make three choices regarding your art career from a marketing and sales perspective:

1. Be a professional full-time artist whose art pays the bills;

2. Be a professional part-time artist who earns extra money from an unrelated capacity to pay the bills;

3. Be an accomplished amateur artist who makes occasional sales.

We'll examine each of these choices with the admonishment there is no wrong choice. They are all right choices dependent on your desires, abilities and circumstances. Don't waste time or breath with anyone who wants to argue the validity of making one of these choices over another. It's not their life or career.

As it should be, the choice will always come down to what is right for you. It can take some serious soul searching on your part to come up with the choice that best suits your situation, but few other choices you'll ever make will have more impact on your career as an artist. It is worth putting in the time to clarify for yourself which choice best suits you.

An artist by definition, according to *www.dictionary.com* is:

> -- One, such as a painter, sculptor or writer, who is able by virtue of imagination and talent or skill to create works of aesthetic value, especially in the fine arts.

Inarguably, the term is much broader than that, but it suffices for our purposes to define those artists who are drawn to read this book.

On Being A Full-time Artist.

Being a full-time artist is a brave career choice. It can be humbling to learn it is a pursuit where the odds are stacked against enjoying great success as a professional full-time artist. It's an area where salaried or hourly jobs virtually are non-existent. Nevertheless, that can also be said of the arts in general, whether seeking to make it on Broadway, feature films or as a writer or a recording artist. For most it is a daunting task to make one's art pay the bills without other means of support.

According to the U.S. Department of Education, National Center for Education Statistics, in the years 1990 - 2002, more than 18,000 Fine Art degrees earned at postsecondary institutions were awarded. This does not account for the tens of thousands of artists who graduated before 1990 or since 2002, or who never achieved a fine art college degree, but whose work at all levels, i.e., full-time, part-time and hobbyist, is actively being marketed and sold — your competition.

Selling visual art is a crowded and competitive field. As in other areas of the arts, success is achieved not by talent alone, but also ambition, luck, selfishness and unforeseen circumstances.

One unique advantage visual artists have over others who strive to make a career in the arts is that they have much more latitude to make their own decisions and drive their own careers if they have the talent, desire and wherewithal to do so.

This book will most help those of you that possess the right stuff and decide developing a print market career is your goal. Perhaps it will help some of you decide that, for whatever reasons, you are missing some elements in the success capability quotient and you should choose not to pursue a career as a print artist, or at least not as a self-published artist.

The good news is if you have the right stuff, you can exercise much more control over your career than artists in other fields. Recording artists, writers and actors are far more reliant on a whole host of people, luck and circumstances to make their careers successful than are visual artists. That is not to say outside influences don't have a great deal to do with the success of visual artists. They do. It's just that you have more ability to create your own situation and make your own luck as a visual artist.

Not to paint (pun intended) too rosy of a picture here. The reality is visual artists also are reliant on powers beyond their control to make their success. In this book you will find stories about successful self-published artists with different looks and styles who share common traits and attributes that have helped to shape their careers and brought them unqualified success, and who, each in their own way, took control of their future and fortune.

A word about success; invariably outsiders measure it in terms of fame and fortune which can be a good measure for many artists as well. Nevertheless, success, as it pertains to artists or anyone for that matter, is truly and wholly a personal definition.

<u>Success is getting what you want from your art, nothing more</u>. The notions and opinions of others with regard to how successful you are, or will be, are not important. Of course, the opinion of collectors, dealers and gallery owners, and of some critics, can be important to your fame and fortune. But none of them can tell you how you should decide what success means to you.

Being a full-time professional artist means you are committing yourself to making a business of your art. As a professional, you have to take actions and make decisions that will influence how you create your art. It's common that some artists upon realizing what it takes decide not to pursue a full-time art career. Some correctly decide they would rather work part-time instead of making compromises on what to create and paying attention to all the other requirements of the business of getting art marketed and sold.

What Is The Difference Between Marketing And Sales?

Many people tend to confuse these terms as being synonymous when they are two different activities. The most basic way of explaining it, and one that will suffice for this book is to say activities that generate interest and action to bring someone to the gallery would be considered marketing. Postcards, advertising, brochures and Websites are all examples. A sale, on the other hand, is the activity when someone is engaged and asked to buy, or when the art is demonstrated and discussed before a buying decision is made.

On Being A Part-time Artist.

It is established that being a full-time artist is a difficult thing to achieve. And, for purposes of this book, I have defined full-time to mean a career in which the artist pays all the bills and earns a living from the fruit of the labor of creating and selling art. If you have a part-time job to make ends meet, even if it's working in a gallery, a frame shop or a museum, you haven't yet met the criteria as a full-time artist.

Being part-time can be pleasurable and satisfying for many artists. If you are one of those people who are multi-talented and enjoy, for example, computer programming by day and creating art by night, you might be lucky enough to have the best of both worlds. And who knows, if your art is the sort that finds a ready audience you may be compelled some day to make computer programming your hobby. You wouldn't be the first artist who crossed over from some other field even if creating an art business was not the original intent.

For many artists, their ideal would be to work at creating magnificent pieces of art that magically were sold at worthy and regularly escalating prices with little or no involvement from them in the actual marketing and sale of their art. As it often goes in life, the ideal and the real do not synchronize.

You'd be hard pressed to find any successful artists that have had no hand in the marketing and sales of their art. And if they currently are successful and aren't involved, you can bet they were involved in getting it marketed and sold before they achieved the success that allows them to concentrate fully on creating new work.

On Being A Serious Hobbyist.

And Is Commercial Success Truly Included In My Vision For Personal Achievement?

Because creating art is such a passion and pleasurable pursuit for many people, the idea of turning it into a business is an abomination and should remain that way for most of them. I am an accomplished fine woodworker and former president of the St. Louis Woodworkers Guild. I used to daydream about turning my hobby into a business and making fine furniture for others. The daydream was as that of many talented would-be professional artists. I would quit my job and create a new occupation by doing woodworking full time. Two separate instances helped to disabuse me of trying to fulfill my fantasy.

The first was a letter to the editor of one of the many woodworking journals I faithfully subscribed to for years. It was a letter to tell all the hobbyists wannabes that they should think hard before they quit their day job. In it, the writer told the story of a woman who came into a professional woodworker's shop and fell in love with a chair on display that he had made.

She was delighted to find out the price was only $100 and she inquired about the price of having three more made to complete a table she already owned. The price for the remaining three chairs was set at $200 each. Naturally, she was surprised and asked why they were double the price of the first one. The simple reply was, "The first was fun to make."

To complete his thought, the woodworker's letter went on to describe a neighbor, an accomplished hobbyist with all the necessary talent and tools to be a professional, who was building a fine cherry chest of drawers. He told how the neighbor had spent six months building it, hand cutting dovetail drawer joints, carved legs and so on. He went on to tell how the neighbor spent another three months completing the piece with painstaking sanding and hand rubbing applications of stain and finish. It would be a masterpiece when done. To the lament of the professional, it was a piece he would never build or sell because he could not charge enough to pay for his time.

The second incident involved my brother-in-law. He is an artist in sheet metal and a master mechanic. I've seen the incredible beautiful result of him cutting apart a 2-door and 4-door 1956 Chevy Bel Air and creating a fabulous convertible out of the pieces. He and his talented brothers have built numerous hot rods and drag cars they race at tracks throughout the Midwest.

After seeing the ability and passion for building hot rods, I suggested to him he should quit his seemingly dull job at the sign making plant and open up a speed shop or a custom car shop to build hot rods for other people as a source of income. He looked at me incredulously for a while and said, "Why would I want to do that? It would ruin my hobby. I work at my job so I can enjoy the time I spend in the garage."

When I heard that and recalled the letter to the editor I'd recently read, it sealed my woodworking career fate, too. I realized the reason I do woodworking is to have a pleasant pursuit with a creative outlet that allows me to design and build objects at the speed I choose — not on someone else's time and budget.

The moral here is...no matter how good you may be as an artist, it's possible that for personal reasons your best choice is to not make it a career. As I'll state later in this chapter there is no shame in success, likewise, there is no shame in not seeking a paying career as an artist. It's a choice only you can make.

For many creative people, one of the most difficult things they have to reconcile is, despite their talent, they aren't going to become a famous artist, musician or what have you. Often situations occur in circumstances out of the control of the artist. Sometimes it is owing to their actions, but it still blocks them from fulfilling their fantasies of stardom in their chosen field.

If it seems as if the opportunity to be on stage in front of a band or the star of a glittering art opening has passed you by, you may need to work on resolving how to overcome the feelings such awareness brings. Perhaps because you are an artist, you are more sensitive to finding yourself in this place with the realization that things didn't go your way.

How you come to grips with these feelings, whether through therapy, faith, spirituality or spending your energy on worthy goals that are within your grasp, can make the difference between being happy or not. Becoming a big star artist with your own publishing company may not be attainable, but if you have the talent, nothing can stop you from using your skill to create art that will live beyond you, and in all likelihood, generate a decent living too.

While you cannot always be able to choose the outcome of your circumstances, you always can choose how you react to them. To understand, to acknowledge your complete control over your ability to choose your reaction, will provide you with tremendous personal power. Realizing your ability to choose your reaction to adverse conditions and to remain positive is how you elevate yourself in even the most difficult situations. Keep that in mind the next time life gets you down.

If the idea of choosing your reaction appeals to you and you want a role model to consider, choose that of Viktor E. Frankl, whose powerful book, *Man's Search For Meaning*, inspired Stephen Covey to write glowingly about it in his bestseller, *The Seven Habits of Highly Effective People.* If Frankl can find ways to choose to be positive during his five years while incarcerated at Auschwitz and other Holocaust internment camps, then by comparison, professional disappointments can be dealt with in ways that, instead of destroying us, make us better people.

Art is a gift to the future and in the future your art will be valued for what it is and not by how much money you made from it in your lifetime. So, the choice to be a serious amateur artist will always be a viable one, even if the path to making decisions about how to apply it for your personal situation has its painful moments.

> What was any art but a mould in which to imprison for a moment the shining, elusive element, which is life itself?
>
> – Willa Cather

Do You Have A Mission Statement?

Having one can be used as your touchstone for helping you make good decisions and to guide you in your career. They work well in helping to guide your personal life too. A well-executed mission statement is dynamic and will change as you pass through stages of your life and career. Nevertheless, its essence will remain constant as those things that define you as a person have remained constant, as you have evolved in all aspects of your life.

If you want some help with creating a mission statement, the free mission builder at the Franklin Covey Website is helpful: *www.franklincovey.com/missionbuilder.* It was developed in part by Stephen Covey, author of *The Seven Habits of Highly Effective People,* which is mentioned earlier in this chapter.

If you haven't read this perennial classic bestseller on self-improvement, I suggest you put it on your must read list. It's not an easy simplistic "Chicken Soup" kind of book to read, but it is positively worth the effort if personal self-improvement and personal leadership are things you want to develop in your life.

What Do I Want To Achieve With My Art?

This question is as personal as what you want to create. It ties directly in with having a mission statement. It is as important to know this as it is knowing what you want to create. You need a road map to get where you are going, or you won't go anywhere.

Imagine yourself being dropped into a big city in another state and you want to get home. Wouldn't a map and a plan to use the map help you get there quicker? Spend all the time you need to think about this question and come up with an answer that is believable and achievable. It will pay dividends beyond your imagination.

Do you envisage yourself a multi-millionaire artist whose editions sell out quickly, or would making enough money to pay the rent and keep the Volkswagen van running make you happy? Would you like to put your kids through college based on your art sales, or does picking up pocket change to pay for a few paint supplies float your boat?

Only you can come up with the answer to what you want to achieve with your art. When you are comfortable with what it is you want from your art, then you can start to tackle the rest of the puzzle insofar as how you are going to turn your goals and dreams into your reality. Many artists are living their dreams. You can be one. First, find your true North Star and then proceed to break your plan down in manageable increments to keep on track with your goals.

If you are the sort who can use the mentoring of an organized goal program customized especially for you and your art career, you owe it to yourself to review the offerings of Artist Career Training: *www.artistcareertraining.com.* Contact director Aletta de Wal. She will be delighted to know I sent you and she will be your personal mentor. I have found her to be an extraordinary person with tremendous energy, passion for her work and great insights into how to make a success from an art career.

I offer Tele-class conference calls through A.C.T. on the subject of getting into the print market. You can either contact A.C.T. or me to find out the schedule for my classes. The feedback I'm hearing about artists who are in the A.C.T. career goals program is that they are experiencing significant improvement in the sales and marketing of their art — something to consider.

Is Being Commercially Viable Part Of My Definition Of Success As An Artist And Will I Have To Make Creative Sacrifices In Order To Achieve Success?

In order for the world around you to know and enjoy your art, you will have to have commercial success. You don't have to sell your soul or make creative sacrifices to enjoy success as you define it. Nevertheless, being able to incorporate suggestions from coaches, agents, mentors, publishers and occasionally even collectors can improve your odds of having greater success.

As with life itself, there are many choices. A choice of what to create and what to do with it once your art is created. It is your choice to decide to pursue art as a career, enjoy it as a serious hobby or find some middle ground between the two. I mentioned how the artist, Arthur Secunda, neatly managed to straddle the poster market without hurting his high end print and collage market. If anything, he felt it gave him and his art greater visibility. I'll talk later about Richard Thompson and how he was able to achieve a similar situation in his career.

Enjoying a career similar to artists, which you admire for the imagery they create and the business model they use can be yours. But, you must work equally hard at making good informed choices about your career as you do about creating the best art you can.

Ambition Is Not A Bad Thing.

Ambition coupled with talent is often the key driver in determining how far someone's career goes. That applies whether you're a salesman, an artist or a bookkeeper. It can make the difference in how a career develops. It can make a huge difference, even at the top, especially in modern art.

Often it is the artist who is the best connected to the right dealers, publicists, journalists and centers of influence that gets the most acclaim. Those things don't just happen; the artist is fully complicit in being involved with those people and schmoozing them in the right way. If you study the most well known fine artists of the 20th Century, whether it is Picasso, Jasper Johns, Robert Rauschenberg and Jackson Pollock or Thomas Kinkade, Bev Doolittle, Terry Redlin and other popular print artists, you won't find any lacking ambition.

There Is No Shame In Being Successful.

Suffering from the high repute, low bank balance or starving artist syndrome is just that—suffering. Despite how fiction and fantasies see this as somehow romantic, it is not a prescription for producing an environment conducive to creating your best art.

Isn't it more shameful to not be successful if you truly believe in your art and crave for it to be known? If few people ever see and appreciate your art, then wouldn't it be a shame to have your creativity go unnoticed when your desire was to share it?

The art world is full of examples of artists who have achieved the model this model of high recognition, high integrity and high sales resulting in a high bank balance. There is nothing to stop you from working to accomplish the same in your career. What you need to do is be able to accurately assess your resources and base your goals and expectations upon them. Use this book to help you begin to take control of your career.

If the artist is true to self and not vying for commercial trade, art is the creative concept taken from the heart of each individual who practices art. It's the formula for being an artist of high repute, and low bank balance. Reality tells us that there has to be some commercial use for the art we produce otherwise it's going to remain unsold and unviewed. OK most commercially successful artists are producing what those who are not so successful call rubbish. But I find that the best driving force is eating, so maybe commercialism makes better artists.

— Batty's Bulletin

Chapter Two

Understanding Art Print Media

> Instead of trying to render what I see before me, I use color in a completely arbitrary way to express myself powerfully.
>
> – Vincent Van Gogh

It is important for clarity's sake to identify the various media in which original works are reproduced as prints — at least those that comprise the bulk of the market of the middle and lower tiers.

The reason to discuss print media here is that the other decisions you will make with regard to self-publishing, or seeking a publisher, are driven to some extent by what kind of art print best suits your work. For instance, if you create novel pieces that incorporate Teddy Bears, you are best suited for the poster market with those images. On the other hand, if you create large abstracts, it's probable you will be drawn to serigraphs or giclées, as your medium of choice, although oversized posters using this genre is a growing market segment.

There are economics around producing and marketing different types of media and typically publishers, who sell posters, don't sell giclées, serigraphs and higher end works. That means you have different decisions to make regarding which medium you believe best represents your art, or best represents your chances for success in publishing, and what companies to prospect for and approach if you are inclined to collaborate with an established publisher.

The following list is not intended to be an encyclopedic or technical explanation of print types, but rather a brief description of those types of prints buyers at ArtExpo and Decor Expo and other fine art tradeshows most often encounter.

Serigraph or Silkscreens

Usually, serigraphs are reproductions from an original, but they also can be an original graphic. In reproductions, either the artist or a person known as a Chromist identifies in minute detail each individual color in a piece of art on a separate screen. As an original graphic, the piece is designed and created as a serigraph.

Although modern materials replace the silk, the screens were originally made from stretched silk, ergo the name silkscreen. With each individual color separated with an opening on its screen to allow ink to be squeegeed onto the art paper in a special press, the printing process begins. Some intricate silkscreens have 100 or more colors meaning that each print must pass through the press 100 times. A run of 350 prints in a limited edition can cost upwards of $50,000 or more to produce, depending on the number of screens in the job.

Serigraphy is one medium that crosses over between top and middle tier shows and is widely used by the printmaking community. Middle tier publishers and artists are also incorporating it in mixed media along with digital printing.

Giclée

This digital print format is the most broadly used form of limited edition print being sold today. It took years for the giclée print to gain full acceptance in the industry because of the controversy surrounding the colorfastness reliability in its early stages, and that it was computer-generated. New inks, solvents, printing processes, substrates and finishes have all but eliminated the criticism. A resistance to technological advances contributed to the hesitancy for many in the art community to embrace the giclée as a legitimate medium.

Since these prints are digitally produced, they can be run in small, even one-off runs, which is the major attraction to them for artists and publishers. It means you don't have to eat your mistakes as you do with other forms of printmaking. By far, the last and best word on the efficacy and worthiness of digital printing is Henry Wilhelm. For all you'll want to know about his extensive research click on: *www.wilhelm-research.com*

If you are looking for giclée printers, it is wise to stick with those who are certified by the Giclée Printers Association – *www.gpa.bz* – or, at least follow the methodology and use the products recommended by the GPA. With the dramatic drop in the cost of equipment, there has been an explosion of shops getting into the business of printing art and calling it giclée. It is critical that you must perform your own due diligence on the companies you would entrust to print your artwork for you.

Limited Edition Reproduction

This is art on paper via the traditional four-color offset printing process, although you see some six-color press runs being used today. It is still favored by many top art publishers to sell offset limited edition prints. Incidentally, the record for largest limited edition print, "Sacred Ground" by Bev Doolittle was produced as an offset print. It has more than 69,000 prints in the edition.

This format is not as dominant as it once was because the only cost effective way to afford an offset print edition is print it all at once. That means you, or your publisher, are betting that enough sales will be made to break even at least on the printing, marketing costs and other overhead notwithstanding. Time-limited pieces as Doolittle's "Sacred Ground" abated this problem, but they are a rare anomaly to the normal process.

The giclée printing process has eroded this market segment for obvious reasons. All the same, on a cost per piece basis, offset limited edition pieces cost less to produce than giclées and still have appeal to some publishers and collectors.

While offset lithographs are not the dominant limited edition medium they once were, they are still used by top rank limited edition publishers as Mill Pond Press, Greenwich Workshop and Hadley House to name a few.

Open Edition Reproductions

The classical definition that has been used by art trade magazine editors is to describe offset lithographic prints as those images printed without any graphic or type on the surface sans the fly spec bold print acknowledging the publisher. Often this format is confused and referred to as posters by the public and even by many within the industry.

Posters

Here the trade magazine editor's classic definition of a poster is that of an offset lithographic image with type or other graphic elements as part of the entire print. When you think of Vintage Posters, those popularized by the likes of Toulouse-Lautrec and other French graphic artists come to mind. That they were used to post the advertisements they bear is where the name poster comes from. This is the distinction between posters and open edition prints in art lexicon.

Of course, Vintage Posters were not printed as offset lithographs, but that is of little concern to the modern poster publisher and consumer. The moniker has stuck with the medium, which in the mass market is always offset lithography.

Now that you have this distinction between open edition prints and posters, I advise you to not waste your effort trying to enlighten others with it — they won't care. Just be satisfied to know for yourself. You will find the word poster used in this book to describe both media since posters and open edition prints are used synonymously throughout the industry.

If you are planning on self-publishing posters or limited edition prints, a good place to start your search for a printer is Color Q Fine Art Printing, *www.colorqinc.com*. They have been printing for artists and publishers for decades and they offer a wide-range of marketing services along with their printing. They can help educate you on the whole process of incorporating offset prints into your career and marketing plans.

Canvas Transfers

Now here is an interesting product. It has been around for decades. Thomas Kinkade is the most famous and prolific artist using canvas transfers as they make up almost all of his limited edition prints. They can be produced for less than giclées and still create results that make it hard to believe they are prints. He uses offset prints to diversify his pricing and offers them more to differentiate product sold through dealers not in his gallery program.

Canvas transfers start out as normal offset prints, and then are covered with a clear acrylic process that binds to the ink. After hardening, the prints are bathed in a chemical and water solution until they soften so the paper can be removed from the bound acrylic and ink top layer. This top layer is slid onto a thin adhesive paper and mounted to canvas using a heated vacuum press. There are other methods, but the above process produces the best results.

The cost to create a canvas transfer and have it stretched on a frame runs from $.50 to $1.50 per united inch. United inches are figured by adding the dimensions of the length and width of the image area to be stretched. Thus, a 16" x 20" image has 36 united inches with a cost of $18.00 - $47.00. Even at the high price, it's a fraction of the cost of a giclée printed on canvas and mounted on stretcher bars.

The effect is something that looks like an original, especially when additional coatings are added to the canvas transfer after it is stretched. Sometimes clear gels are applied to give the effect of brushstrokes on the pieces. Other times, paint highlights are added. Kinkade even had a van tour of galleries with his highlighters featured...go figure. An apparent advantage is that, unlike other prints, canvas transfers can be displayed without using glass to protect them. This way, they show great in galleries and on the walls of collectors.

Now that you are well informed with some basics on the different kinds of prints that are most often seen in the marketplace, you are getting a clearer idea of which of these methods best suits your own art and needs. By coming to grips with the question of what medium best suits you and your art, you can progress to other decisions regarding whether you should self-publish or seek a publisher. Should you decide on the latter, with this decision made, you will be better prepared to target those publishers who use the medium you have chosen as ideal for your art.

Much more could be said about each of these and various other art print processes. The intent here is not to burden you with more than you need or want to know, but to give you some guidelines and basic understanding of the overall commercial art market from a print medium perspective.

As to decisions regarding being a full-time or part-time artist, there is no wrong choice here. It often comes down to economics, subject matter, personal preferences and competition to help you make the decision. Naturally, if you have inputs from a publisher, it will help influence your decision.

Later on I'll provide some ways for you to spot trends in current art themes. Keep in mind at the same time you are sleuthing for ideas on trends, you also can put print types and which publishers are using which formats into the mix of things that you are tracking. This also will help you make sense of the business on multiple levels.

More On Giclées

The format warrants further discussion as it has become the medium of choice for the majority of self-published artists and middle tier publishers who sell at price points above posters and offset limited edition reproductions. Giclées offer too many advantages against only a few drawbacks for them not to have become the important part of the industry mix they are. Spend some time at ArtExpo and other premier middle tier shows and you'll note the prevalence of giclées.

As I have noted, the format is an advantage for publishers because it eliminates expensive print runs for untested images. That reduces overhead for stocking and expenses for inventory not selling. It also allows publishers to sell images at different sizes. A trend in the last few years has been in multi-size editions. That is, the edition may be limited to 350, 500 or 999, but the image sizes themselves may be different. As long as the aspect ratio stays the same and it fits the customers' needs, the thinking is to give the customers what they want.

There is a lot of flexibility with giclées as you can see in the trend for selling complete images in different sizes (multi-size editions) and selling one edition of canvas giclées and another on paper, some editions in different sizes, but all with the same image. This can be confusing to the collector and denigrate the artist's sales in some cases. Apparently, it is not slowing down sales for top selling artists. Check out the Website of Cao Yong to see how he handles it: *www.caoyongeditions.com*.

Another practice used by some publishers is to sell Artist Proofs as a separate edition at higher prices. This is interesting since the original concept of an A/P was that they were the first pieces off the press and presumably, the printing plates were printing at their crispest meaning the fidelity was higher. Nevertheless, with digital printing, ostensibly Number 10,000 should look as good as Number One.

Some artists do small pencil remarques on the margin for some of these "special" A/P prints or add touches in other ways. A/Ps are used more as a marketing ploy now, but as long as collectors enjoy buying them, who is to argue? How you handle A/Ps is one of the many things you have to grapple with as you work your way towards publishing stardom. A/Ps are the most popular, but if you look around you will see other designations and creative numbering methods for edition extensions.

As with the multi-size editions and so forth, the more you move away from an easy to understand approach to numbering your art, the more it is possible you are going to spend time explaining, and in some cases even defending, your edition numbering system. If you are just getting started, keep it simple for your own sake and that of your collectors and galleries. If you find yourself with so much demand for your work that you are selling editions in several formats and sizes, then you should be grateful for your good fortune. Just don't let your success ruin your success.

Artists are fond of the giclée format as well. They like the broader color spectrum it offers, especially on the newer 6-color presses. It speeds up the proofing process and the time to make changes is compressed too. It allows them lower upfront costs whether they are self-publishing or not. And, since there is less spent to launch the edition, more money can be put into creating more images. Think about how many different images a publishing company can come up with in the giclée format instead of dropping upwards of $50,000 or more for a single serigraph and you understand how compelling the medium has become for artists and publishers alike.

Collectors have a stake in the giclée format as well, although one can say it was never consumer demand that brought the giclée to the forefront of the business. Nonetheless, collectors can enjoy purchasing a giclée made to order for size, something that never was available in the print medium before. The discerning can also appreciate the improved color gamut.

While not being aware of it because it happens in the background, consumers benefit from the giclée format in two ways. Images are available in different sizes, and there are more total images available. When you as an artist don't have to bet the ranch that the one you are going to press with will be a winner, you can broaden your line. Some could turn that argument around to say that too many poor quality images are available, but I believe the cream rises to the top. Publishers still struggle with figuring out what is going to work and why. When they do, having more choices makes things easier for them and their buyers.

Many consumers are thrilled that images printed on canvas minimize the glare and extra framing costs of matboards and glass. On the other hand, some are resistant to the price points because they see the format as digital and have difficulty equating the value of the piece when thinking of it as digital print.

A note of historical interest is provided here: The word giclée is French and is a feminine noun mean squirting or spraying; hence the spraying of inks or dyes on paper or canvas. The legend of the origin of the word goes back to ArtExpo circa late 1980s when some of the format's pioneers gathered to exhibit at the show.

Realizing then, as now, that the words "digital print" lacked sufficient sex and marketing appeal, they wanted to come up with a word that better described this new process. In what was a stroke of genius and luck, Jack Duganne, using an English to French dictionary, came up with the word *giclée.* From that instance a word was coined, an industry was created and history was made as everyone at once agreed to begin using the term to describe their process.

There are other versions of this story, but the attribution of the genesis of the name always goes to Duganne. And printmaking folklore aside, the term has stuck and become part of the art lexicon.

Art is long, life is short. — Ars longa, vita brevis.

– Hippocrates

Chapter Three
First Things First

It's not what you see that is art. Art is the gap.

– Marcel Duchamp

It may seem odd to have Chapter Three labeled as First Things First. But the first two chapters were put in place to get us here where we can start asking the most important questions about this quest to break into the print market. It is with this chapter that I intend to get you to begin to think about organizing into a hierarchy now what is important and urgent for you to begin to launch your print market career.

Homage is paid here once again to the brilliant Stephen Covey as the title for this chapter is borrowed from Habit Three, "First Things First" in his enormously influential and highly recommended bestseller that remained on the charts for more than a decade, *"The Seven Habits of Highly Effective People."* That Covey feels strongly about this habit should be obvious. He has since devoted an entire book to the subject, aptly titled, *"First Things First."* He uses this phrase to broadly talk about personal leadership and how, if you don't tackle that which is the main thing first, you are dooming yourself to wasting time if not outright failure.

For our purposes, we will narrow the concept down to deciding where your art fits into the market and what actions you should take to support that decision. Getting into the print market requires the 3 "D's" — Desire, Discipline and attention to Details.

Having your own effective plan will help you implement the 3 "D's." The plan will help you avoid making mistakes, wasting time and money and give you the best chance for success as you embark on the path to profiting from the art print market.

Answering these questions will help you get First Things First:

- How can I tell if my art is appropriate for being published?
- Are there reasons to not attempt to break into the print market?
- What media best suits my art?
- Should I be thematic or show a range of work?
- Should I self-publish or seek a publisher?
- Can I do both?
- Is there another alternative?

How Can I Tell If My Art Is Appropriate For Being Published?

You probably already know without asking if your art is appropriate for being published. It's likely why you are now reading this book. Nonetheless, here are some "ifs" to consider regarding whether your art is right for the print market.

- If your originals are getting prices that take too many collectors out of your buying pool.
- If your art is in such demand that you could have sold your originals many times over.

- If you can't produce on a schedule to meet the demand for your originals.
- If you desire or need a secondary source of income from your art.
- If you are creating art specifically for the print market.

Did you have affirmative responses to these questions? If so, you are a great candidate to jump on the print bandwagon.

Are There Reasons To Not Attempt To Break Into The Print Market?

Yes, and they include:

- That your subject matter is too esoteric or inappropriate in that it falls out of mainstream tastes to warrant going into prints.
- If your original work is not moving, you may compound your situation by investing in the print market. If this is your case, most publishers will grasp this whether you tell them or not because it is their job to have a finger on the pulse of the market. That's not to say you would never get a bite from a publisher, but it's far less likely if your originals aren't already selling well.
- Or, it could simply be that after investigating the market and what it takes to be in it, you simply decide not to pursue publishing.

If reading this book helps you make the decision not to pursue publishing, that is as good a reason for me to have written it, just as it would be to have convinced another artist to take the print plunge. The idea is to help stimulate your thinking and deepen your knowledge so you come up with the right conclusions and act on them appropriately. You don't want to have regrets later on.

What Media Best Suits My Art?

Deciding what media best suits your art can be tricky. I mentioned Teddy Bears earlier. If you are painting them or similar content, open edition prints or posters are where your art will find greatest acceptance. Red Skelton, the famous comedian, was also well known for his clown images that were limited edition prints. It was his celebrity allowed him to move beyond posters with this genre. Without it, the art would have had far less appeal to collectors.

It takes some thinking and investigating on your part to decide where you think your art fits in the pantheon. There are price points all over the board. You can be in inexpensive limited edition prints, medium priced giclées or sky-high priced serigraphs or the inverse of any of those scenarios and still make a right decision. You have to decide what feels right for you and listen to what the market is telling you. You may get feedback from collectors and publishers that differs with your own perceptions of where your art fits in the marketplace. If you do, pay attention.

I strongly recommend a field trip to ArtExpo New York, *www.ArtExpos.com.* It is the Mecca of the art print industry and is held in late February or early March each year. If you have never been, you will be in for a sensory overload delight that will alternately scare and excite you with the possibilities and the depth of the talent and competition on display there.

For more than two decades, ArtExpo has been at the center of the contemporary art print business. Just about anyone who is anybody will be there with a booth to show the industry his or her latest images. It's a wonderful and fun way to gain a quick education on how those in the industry operate, display, market and sell their work.

If you can't make a trip to ArtExpo New York, then as your second choice, shoot for ArtExpo Atlanta, held annually in September. Attending one of these shows is not necessary, but it's certainly a great way to jumpstart your knowledge, motivation and inspiration.

Yes, there is an annual art print fair at the Armory in Manhattan. The International Fine Print Dealers Association produces it each year, along with another called "Works on Paper" in Manhattan, both in the spring. Don't be confused, the works at these shows are "elite prints", often old, rare, and expensive and highly collectible and almost all would be considered top tier art as opposed to contemporary middle tier fine art at shows, including ArtExpo.

Breaking into one of these top tier shows is something to aspire to for some of you reading this, but it's likely that by virtue of reading this book you are a few levels away from participating at them. But also, by virtue of reading this book, you are educating yourself on what is possible if you keep progressing in your career.

Should I Be Thematic Or Show A Range Of Work?

Many artists are talented in that they can paint a variety of subjects and themes equally well, and enjoy stretching themselves to try new things. Having a roving eye is not the best way to make a career in the print market. The way to success for artists in the market is to become known for something recognizable. If I say cottages, unless you've been living in the proverbial cave, you're going to reply, Kinkade. LeRoi Neiman is known for his colorful sports images and P. Buckley Moss for her quaint Amish scenes and so on.

The point here is, it's really about growing a collector base and keeping your dealer base happy knowing you are hard at work creating more images for the buyers they market to in their areas. It's not to say you can't paint what you want to paint, but if you are making a business of it, you need to establish a look and style that is uniquely yours and stay with it long enough to become recognized for it.

Alternatively, you can look for a publisher who will feed you ideas on what is needed to meet developing trends, but it becomes much more of a hit and miss proposition for all involved and the probability of finding a publisher who would take you on in that kind of capacity is small. So, pick something and stay focused on it. If, after a while it's obvious that the fish aren't biting, consider changing your style.

Keep in mind some poster publishers are perfectly fine with the idea of you painting under an assumed name, a "Nom de Brusse," if you will. Most aren't looking for you to make personal appearances for them anywhere, so there is the possibility you can create a look for a poster line that is not something you want to be known for with your original collector base. Think of it as moonlighting, or guilty pleasures. Alternatively, keep in mind the previously mentioned Arthur Secunda and realize you also could model after him.

Should I Self-Publish Or Seek A Publisher?

This is the $64,000 question that got me started counseling so many artists during my 15 years with *DECOR* magazine: Should I self-publish? My short answer was then and now is, "Yes, if you can." It may be a short answer, but that "IF" in there is a big, big "IF." Frankly, for most artists, it's not really a viable choice. This question, more than any other, is really getting to **First Things First.**

I say, "Yes," because if you can successfully self-publish, you should because the rewards are greater and the control of your artistic life remains in your hands. Also, if for whatever reason, you should decide that self-publishing is not the best thing for you, having done it will give you a greater appreciation for what a publisher does when you decide to throw in with one.

At this point, you will not have read enough of this book to make the decision about self-publishing now, but it will need to be one of the first decisions you make when you seriously consider putting your work into print. We'll cover many more aspects of self-publishing, including many examples of successful self-publishers, in later chapters.

Odds are almost all of you will end up going into print with an established publisher. I say this based on my experience and by judging the number of successful self-published artists versus those working with publishers. As with so many other decisions, this isn't about being wrong or that one choice is better than another. It's really about making the right choice for your situation. If you have a full-time job to pay the bills and no real unpaid support help that allows you lots of time and energy, self-publishing is going to be tricky if not flat out impossible for you.

On the other hand, if you have help and are already making substantial income from your art, you are a good candidate to move forward with a self-publishing business plan. In subsequent chapters, we'll look at examples of self-published artists and successful artists who work through professional art publishers, along with suggestions and ideas on developing a marketing plan for each scenario. We are talking about these things now as a way to get you thinking about them as you move through this book and progress towards making career decisions regarding getting published.

Can I Do Both?

Some of you will want to try to have a publisher and do some self-publishing as well. That is perfectly fine so long as your publisher doesn't mind. Some are open; others want to be restrictive with you on where else you can market your work. It's something you'll need to investigate and keep in mind as you begin contacting publishers.

Some artists license their images as posters to publishers while marketing other images in other formats on their own. This is a typical situation for many artists whose creative output cannot be handled by their publisher. You have to work out what is best for you and your publisher.

Remember to ask many questions when you negotiate with a publisher. (There is a whole chapter devoted to working with a publisher.) Don't be intimidated and do stick up for yourself. You have a right to make a living. If a publisher only wants to cherry-pick your images, then you can't allow yourself to be tied up to some exclusivity unless you really want to be a starving artist.

Is There An Alternative?

There is another alternative to publishing, especially self-publishing. We'll get into the economics of self-publishing soon enough. For now though, let's say that for a fraction of what it will cost you to launch a self-published art business, you could take the vacation of a lifetime.

I've cringed more times than I care to recall after receiving a call like the one I received once from an artist who had borrowed money from her parents to make a couple of limited edition prints and now she was asking about buying some ads to promote them using a credit card. We're talking in this instance about committing thousands of dollars to try to launch a career with two prints and a few full-page ads in *DECOR* with not even enough money to get into a tradeshow. In my professional opinion, this is a prescription for disaster.

My advice to that artist, and to you now if this even vaguely describes your situation, was to take the money and go on the vacation of a lifetime instead. Make it so exotic and extraordinary that you will talk about it the rest of your life because if you choose to go into self-publishing unprepared and under-funded, you are, as they say at the poker table, dead money. If you take this advice and go on vacation, don't forget to send me a postcard and toast me while watching the sunset in Bora Bora, or whatever exotic locale you choose.

Okay, we've described the top-level opportunities in publishing and briefly discussed the merits and pitfalls of each. You're still a long way from having all the information you need to know about self-publishing and using a publisher for you to be able to make a decision about which is best for you.

Nevertheless, by reading this far you're getting closer and you have established what is first in importance on the decision tree (self-publishing versus seeking a publisher) and you are beginning to gain some insight on what other downline decisions you'll need to make to get your print career on track. In other words, you have your attention focused on **First Things First**.

Chapter Four

Traits and Attributes of Self-published Artists

> Those who do not want to imitate anything, produce nothing.
>
> – Salvador Dali

In every endeavor where one wishes to succeed and indeed, exceed those masters who have come before them, studying the tools and techniques of those masters is the basic prescribed method of improvement. Learn everything you can about how they did it and do it better. It is human nature to push for faster, stronger and more beautiful and consequential. We have intelligence and we use it to strive to better our surroundings and ourselves. Studying successful self-published artists to discover what they have done to gain their triumphs is natural and logical.

Study the careers of successful self-published artists and certain traits and common attributes become obvious. The list that follows is arguable in that other items could be added to it, but there is no argument that each item below is critical to successfully launch and sustain a self-published art career:

- Talent
- Art that resonates with a large group of collectors
- Financing
- Personnel, usually a spouse, devoted family member or close friend

- Willingness to prodigiously continue to create art that is in the same thematic range in order to continue to supply the dealer and collector base

- Ambition

If with no further clarification on the above points you can honestly evaluate and see that you possess the qualities and capabilities outlined, you are a prime candidate for a top-notch self-published art career. If you have taken a frank evaluation and find you are lacking in even one of these critical areas, you probably should consider working with an established publisher, at least until you are confident that you fit the criteria above.

If that sounds blunt, you're right, it is. It is intended to be. It's a way to set the tone so that the following statement is clearly understood:

Never forget that although the art business enjoys a certain glamorous *je ne sais quoi* quality about it, which depends on you tapping your creative well to bring forth your artistry to create visual pleasures with deep meaning, it is still a business and must be run like one.

There may be rare cases where an artist remains isolated from the business aspects and is successful, but huge odds are stacked against any who initially would try that route whilst secretly harboring the wish for success.

Talent

By reading this book, you should be coming to a self-assessment in many ways, not the least of which is talent. Unfortunately, many talented artists do not enjoy the commercial success of their less talented contemporaries, and not just in the visual arts. Why else would the pop charts be filled with pretty faces sporting reed thin voices, why would some artists that critics dismiss as hacks make fortunes? Because hitting the big time in any field takes more than talent, especially in the arts that are so subjective and because the level of sophistication of the mass consumer is low. That is not to short change talent. Nevertheless, if you lack it, your road will be rocky.

Talent within themselves for some people is hard to gauge. In the visual arts, it is subjective, and thus leaves open the question for some artists with a lesser talent to delude themselves into thinking there is more there than reality. Often, these artists are spurred on by the best intentions of family and friends who lack the critical facility to give them honest advice on their capabilities.

For some whose aspirations are larger than their talent, it can be painful to come to the realization that they fall short in this category. All the same, if they draw an accurate bead on their ability, they are likely to avoid even further pain in the form of financial setbacks. Bottom line...get critical opinions from qualified but otherwise uninterested parties.

I often saw that artists with far too little talent or development would optimistically come into the gallery where I worked, or into the booth of a publisher at a trade show, hoping to find acceptance. Instead, what they typically found was polite rejection. If you find yourself in such situations and have the courage, you can ask for a quick appraisal with the caveat that you would deeply appreciate an unvarnished estimation.

In some areas, sports for instance, it's harder to hide a lack of talent. If you can't hit a curve ball, you won't make it to the big leagues, no matter how good an athlete you are otherwise. Look at Michael Jordan who strived for two years hoping to ignite a second career in baseball after simply conquering basketball. That it's not that easy, as Jordan can attest, would be the understatement; he gave up and went back to glory and more NBA titles.

What does that mean to you as an artist? That talent matters, but that you can make up for some lack of it if in other areas you can excel in other areas. This is not to say that you can be really bad and successful, but there is plenty of evidence that those artists with less than stellar ability and who do far below museum quality work make a great living selling visual art.

The bottom line is if you have talent, you can learn technique and make a go of it. If you lack talent, your ambition and business sense will need to be off the charts to get there. Taking an honest, brutal assessment of your own talent is often hard to do. But it can help you adjust what you need to do on the side of imagery, content and style. Don't take on more than you can handle; it will only add to your frustration. The sports cliché for that is, "Staying within yourself."

Art That Resonates With A Large Group Of Collectors

If you can see the fish and they aren't biting, then you are using the wrong bait. That buyers must like your art may seem obvious and I hope that it was for you before reading it here. Possessing that knowledge is proof to yourself that you are conceptually on board with what it takes to make a print career.

Your business is really about supplying dealers and gallery owners with your art. Not surprisingly, dealers enjoy developing collectors who will come back and buy more art of a nature similar to what they first bought. If you are doing well in landscapes and switch to painting trains, good luck. Your collector and dealer bases have now moved back to square one.

Thomas Kinkade has built a $100 million dollar business around painting cottages and nostalgic scenes of homes and beautiful romantic vignettes, *www.thomaskinkade.com*. I can imagine some days when he sits to paint cottage number (pick one) that it requires great discipline to do the work and it is a chore as much a labor of love at times.

I don't know the man, so this is only an impression on my part and for the sake of his dealers and collectors I'm sure he would never admit to a lack of desire to paint cottages and anyway that's beside the point. Regardless of whether you admire the man for his success or vilify him for it, you cannot argue he is the most successful limited edition artist of all time. Whatever his motivation, he personifies the 3 "D's" (Desire, Discipline, Details) and continues to crank out the product.

His achievements were made to a great degree by finding a subject matter and style that has compelling appeal for a large group of collectors and continuing to, in a manner of speaking, feed the beast. Something for which Kinkade rarely receives credit is that he has created a large body of art collectors from those who would have otherwise never set foot in a gallery. Some of them have grown in sophistication and look for better art — maybe yours.

Compare the art of P. Buckley Moss, *www.pbuckleymoss.com,* and her nostalgic scenes with Wyland, *www.wyland.com,* and his ocean art and you will note each built their business selling to different kinds of collectors by finding a theme and style that resonated with them and then continued to layer on more of the same year after year.

These are but three of innumerable self-published artists who have found and kept success by developing a subject matter, look and style that collectors enjoy and then have continued to supply them with more of the same. You will need to consider how you can emulate developing a look and style of painting for your own business. It is a key component of the whole picture.

> Repetition is the mother of all skill.
>
> – Tony Robbins

Financing

The adage, "It takes money to make money." is as true in the art print business as it is anywhere else. This is where having the intangible of luck can play a crucial part in the development of your business. That is, if you already have money, you can self-finance, or if you have access to someone who believes in you and who has money, you are in luck. If you are like most people, having excess funds available to throw at a risky business venture such as art print publishing is not a reality, you need to be creative in a whole new way — that would be in raising money.

The number one reason why small business startups fail is lack of adequate funding to keep them afloat while they find footing and develop a client base. It is the same, maybe even more so for self-published artists. Brilliance is not required to say among all the attributes and traits that cause would-be self-published artists to fail or not even try, lack of financing is the number one reason why.

You might recall in an earlier chapter I remarked that almost all readers of this book are better candidates for seeking a publisher than for self-publishing and inadequate financing is why. I'm not in the business of dashing hopes, but I'd rather be known for that and save some of you from problems you never imagined possible by laying out the brutal honest truth of what you might be considering than cheerleading you into financial ruin. Having a garage full of unsold paper, crushing debt and broken optimism is a much harsher reality than admitting you don't have sufficient funds to get started self-publishing.

I'm sure for those of you who are already on the way and those of you ready to storm the gates; these admonishments will mean little, if anything. That's okay; they are not really intended to slow your roll. Of course, my fondest hopes are that almost all readers will come away with a clearer idea of how to proceed and follow those ideas with appropriate action that leads to success. I'd love to hear about it. Write me when you get there.

Regarding financing, there are no tricks to tell you. Among artists who are already successful, they hardly ever discuss their early financing adventures, so anecdotes are rare and not readily available about how they did it. Some had luck with wealthy family members, but reality says more found luck in the residue of hard work. That is to say, they built their businesses little by little in the early stages and continued to invest earnings back into the business. Others hit it right with a look that defined a trend or created one and caught fire from inception.

There is an abundance of successful self-published artists not well known outside their dealer and collector base. They are not media stars or publicity hounds. Yet, they have fabulous careers and enjoy nice incomes from their art.

Sometimes I think they have the better of it just as I think of those fine actors who work the stage, not on

Broadway but at places as the better Repertory Theatres found in large cities across the country. They are well paid, get to practice their craft to the best of their ability, travel to new cities and reside there for a time and otherwise enjoy obscurity rather than fame.

There is nothing wrong with a career like that in my book. It also could be said for symphony musicians and so on...must be good work if you can get it and sustain it. The same might be said for being a fine major league ball player that is paid well and respected by his teammates and yet is able to avoid the hot lights of publicity. He takes pleasure playing the game he is good at, but enjoys a degree of anonymity that the better-known players gave up long ago.

Anecdotally, many of those unheralded yet successful self-published artists started slowly and steadily grew their business. First at local shows and hustling like everyone else, getting attention any way they could to make sales. Then they moved on to regional shows and juried affairs, all the while studying the moves of those whom they admired that were already enjoying self-published success and improving their art and marketing skills at the same time. Finally, amassing enough money to beg, borrow and steal the funds to start printing and advertising and attending Decor Expo, ArtExpo and other tradeshows.

There are self-published artists who started out working for an established publisher, but for reasons too different and numerous to mention used their success and reputation to go on and start their own operations. The results are as varied as the reasons they left the relative security of working for publishers. Those with the rest of the package of traits and attributes as outlined at the beginning of this chapter have done well. Those who bolted but without good business skills or a plan or lacked financing, or proper support and so forth, struggled and some outright failed only to end up back with a publisher.

To get there you need to already have money (lucky you), or you borrow it or you grow into a position of having a cash flow to pay for your publishing business. I don't know of other ways to do it. Some combination or permutation of these three methods is possible, but you have to have access to cash to get your self-publishing business off the ground.

Personnel, Usually A Spouse, Devoted Family Member Or Close Friend

Here again, we are talking luck of the draw. If your spouse has a career he or she enjoys, or lacks management and marketing skills, or is up to his or her ears running a household with three kids, or what ever, and your sister, brother, best friend and other candidates are likewise involved, I'm sorry you drew a bad lot on this attribute. Your hill is going to be steeper to the top of the successful self-published artist heap.

The history of successful self-published artists is filled with examples of artists who had the good fortune to have someone on their side that believed in them 100% and who was equally capable of helping them in the critical areas of marketing and management. Lacking that asset has probably killed off nearly as many potentially successful self-published artist careers as has lack of financing. These two attributes go together hand in glove.

Hard as you might try, it's really close to impossible to pull it off yourself. How are you alone going to create the images, raise the money, market the work and manage the business? Not to say it hasn't been done and hats off to those who have, but only the rare can do this.

Such a person, who though mortal, is one who is vastly endowed with left and right brain capabilities, is highly organized, keenly competitive and is an ambitious and talented painter with an eye for what collectors want. He or she would be equally adept at handling marketing, sales, promotion and management duties while still finding time for a life, much less sleep. That is a tall order indeed.

"Monomaniac on a mission" is the apropos quote I borrow from Tom Peters in his bestselling business book, co-authored with Bob Waterman, *In Search of Excellence: Lessons from Americas Best Run Companies.* It best describes many of those whose support has helped drive the success of the art industry's best-known self-published artists. A prime example would be Pat Buckley Moss. She has derived much of her success from the driven marketing leadership of her husband. The duo worked as hard as any self-published artist team ever has at building their business through their dealer network. The success they have enjoyed as a result is tremendous. The notable 20th Century Impressionist, Richard Thompson, had his son and daughter-in-law as his business partners. His is another example of the built-in marketing management attribute.

Another interesting case in point is Marty Bell, *www.martybell.com.* She and her husband, Steve, built a fabulous business around her art, including a devoted collector society, something she shares in common with Pat Buckley Moss. It's been said when Thomas Kinkade looked around for an idea on how to harness his artistic talent that he decided to emulate the cottage look that Marty Bell had begun creating years earlier. He, as

Marty, used his images to build a loyal collector base or society, as many publishers call them.

You can't get a much better example of a superb result from collaboration than that of P. Buckley Moss and her devoted spouse than to see Pat's picture in the center of the cover of "*Parade,*" the Sunday supplement that is delivered to tens of millions of households each weekend. The story she was part of was a feature *Parade* does from time to time called, "What People Earn."

In the article, readers found P. Buckley Moss with her thumbnail headshot among dozens of pictures of millionaire basketball players and everyday folks as policemen and librarians, truck drivers and schoolteachers. Each had their annual salary listed below their names and under Pat's was the amount of $600,000. Now that was back in the mid-1990s, but still not a bad annual take for a painter who favors Amish children, country life and geese.

We'll never know if that was her personal income and if it included that of her husband, although my guess is it was not inclusive of his income or cash that flows to her in other ways. It doesn't matter because that is a greater figure than many reading this book can imagine as a personal income. A half million-plus annually is to truly be in clover.

The point is there are artists making lots of money out there and the majority of them have trusted partners, often family members, who are instrumental in growing and supporting their businesses. It's difficult to recruit a willing person with the right qualifications if you don't already know one. Often they will need to work on the promises of things to come with little or no income at the outset.

It's a cruel fact that it's near impossible to do it alone in the art publishing business. So, if you have that marketing maven, that "*Mono maniac on a mission*" targeted or better, recruited, then bless you. You've filled an important component of the successful self-publishing needs matrix.

Willingness To Continue To Thematically Create Art In Order To Supply Collector Interest And Your Dealer Base

This attribute was discussed earlier about Thomas Kinkade when I mentioned his great discipline in sticking to what makes him successful. That is not to say he doesn't yearn for acceptance in other areas. For example, he has a line of plein-air images that are, in my estimation, a refreshing departure from and better than his cottages and nostalgic work. Nonetheless, his collectors aren't wild for the stuff, and while it gives him a creative outlet for something besides cottages and so forth. I believe if he were trying to become a multi-millionaire based on his plein-air style, he would be up the proverbial creek without a paintbrush or paddle.

The reality, which can be a harsh taskmaster for some artists — even the multi-millionaire artist — is if they are going to continue to feed the beast they must continue to paint in a style and thematic range that is comfortable to their collector base.

The same is true of other artists. Imagine the sublime blues-recording artist Eric Clapton attempting to create a collector base with a rap album. The more success you have the more people in your supply chain you have depending on you to continue to paint those cottages or what have you. You've got employees, dealers, printers and even magazine and tradeshow reps, as was formerly the case with yours truly, hoping and pulling for you that you continue with zest to create and market art that sells at ever increasing speed and prices.

For some artists, this is not a burden, but a joy. There are those who simply enjoy creating images in the same vein, who aren't bored with yet another marine wildlife painting, or Impressionistic landscape and so on. To those, I salute you for your ability and say you should be relieved to not carry around the weight of feeling penned in by circumstances and success. To those of you who feel the weight and succeed despite it, may the blessings of your success help lift you and otherwise make your life so sweet that you carry your weight without grudge or gripe.

Ambition

If you lack this, skip to the chapters on seeking a publisher. No joke. It's that simple. Either you possess it like you possess artistic talent, or you don't. Ambition is not a technique you can study to improve upon. It's either or it's not — simple as that. It's an innate trait built into your DNA. Yes, we all have it to some extent, but you know what I'm talking about — the burning desire to enjoy success and be somebody come hell or high water. Ambition and competitiveness are close allies in success. I earlier mentioned selfishness as an attribute that contributes to success and I believe it is a component of ambition.

The aforementioned Michael Jordan is a case in point. Like all pro ball players, he spent decades traveling around the country for six months at a time to play ball and more time to tend to public appearances and promotional activities. Mark Twain was known to spend enormous amounts of time away from his family in his writing studio or traveling to speak and for other reasons. How many visual artists obsess on their work? Too many to count or list.

For many, they are driven beyond the average person's capacity to understand the process or even to understand it themselves. There are plenty of examples of artists and other high achievers who had to sacrifice something to make their goals. P. Buckley Moss making 100 one-person gallery shows annually for 15 years had to have made personal sacrifices to adhere to a demanding schedule. To wit, she learned to paint on airplanes to best use her time.

Ambition, like talent, comes in different depths in each of us. If you are fortunate enough to understand your own, you can use that knowledge to help harness your success or at least accurately gauge how far you can or want to go with it. If you did use the Mission Builder program mentioned in Chapter One, perhaps it helped you gain some insight into your own makeup and drive.

Let's look back once more at Michael Jordan because his achievements are legendary. He is known to be a fierce competitor who would never accept defeat without giving it his all whether playing a "friendly" game of ping-pong or card game in the locker room. His desire to win and be the best inspired his teammates to be better than they would have been on their own. Those attributes combine into leadership, both personal and organizationally. As the driving force in your business, your competitiveness, ambition and leadership abilities will make a huge difference in your success.

When deciding whether to self-publish or seek a publisher, to be a full-time, part-time or amateur painter, there is no shame in making an honest evaluation and concluding you don't have the fire in the belly to go through the gyrations to gain all the sales and success that you might because you could. Remember the woodworker's story. It applies here as well. You get to define what success is for you.

When I sold advertising for *DECOR*, I had one artist who infrequently advertised. When he did, he invariably pulled the best reader response of any ad in the issue. After noting this occur for sometime I visited him and encouraged him, much like my brother-in-law, the hot rod builder, (Are you getting the impression of a slow learner or a dogged optimist here?) to embrace his success and advertise monthly and really push his sales envelope.

He explained to me that having the ability was not a good enough reason for him to do it. He already knew he could have substantially greater sales by being more visible, but he didn't want the headaches that went with it...more staff, more shipping, more printing, more this and that. Go figure because I had others advertising with me who would have done anything to get his results and would have eagerly pursued every possible sales opportunity if they were open to them. Sadly, they were just getting by and couldn't afford the advertising I envisioned for the artist mentioned here. Life's twists are strange, cruel and hard to comprehend at times.

If you are not driven, passionate and ambitious about success with your art career as you are about creating your art, if not more so, then self-publishing is likely not your cup of tea. It requires drive and much personal leadership to succeed as an art entrepreneur and ambition is a key ingredient in that success.

Do yourself an enormous favor, because it is vitally important for your career path choice. Be honest with yourself regarding your depth of ambition as you contemplate your career as a self-published artist. If your actions and deep desires don't match your wishes, your self-publishing dreams are in jeopardy.

> Fortune favors the bold.
>
> – Basil King

Chapter Five
Economics of Self-publishing

> The holy passion of friendship is so sweet and steady and loyal and enduring in nature that it will last through a whole lifetime, if not asked to lend money.
>
> – Mark Twain

This chapter is not intended to cover all aspects of finance as it relates to a successful art career. There is enough in that subject for a complete book. The objective here is to inform readers with some of the typical costs involved that are specific to the topic of art publishing and to provide some basic formulas for calculating your launch prospects.

The items I'll mention will cover some of the financial variables involved with starting and operating a self-publishing enterprise for artists. To try to be inclusive is not possible as there are too many variables for them to be fairly applied in general terms.

Trying to get one's arms around all the variables necessary to make blanket statements about what it would cost to get an art-publishing career off the ground is a daunting task. When pushed for an answer, I usually apply the S.W.A.G. Factor (Alternate **Scientific** with **Stupid** before **Wild Ass Guess** for the acronym depending on how well you like the answer) and toss out $100,000.

That number is for an aggressive rollout campaign waged over the course of a year or eighteen months, where one could easily spends upwards of $100,000 in getting a publishing career launched. And even that seemingly high figure does not include all the attendant costs in making the operation run. The best scenario is one where things take off fast and the business self-funds most of the costs, but don't bet on it.

> When you have them by the wallet, their hearts and minds will soon follow.
>
> – Japanese proverb

How Then Does $100,000 Get Spent?

6-pages trade advertising	$18,000
Tradeshow space and costs	$30,000
Publicity	$6,000
Printing Art	$12,000
Promotional Materials	$2,000
Business Expenses	$1,500
Shipping Expenses	$3,000
Salaries & Benefits	$45,000
Total	$119,000

Obviously, there is much latitude in each of these numbers as there is no way to assume that every artist jumping into self-publishing will produce the same number of images to be printed, sell them at the same rate and price, have someone working full-time later, but working unpaid in the early going and so forth. There is no end to the different number of scenarios that actually can arise as you begin to put together your plan and your budget. It will end up being what it is and what you make it.

While the other items mentioned above are straightforward, the $30,000 for tradeshow expense perhaps needs some clarification for those who have not gone through the process. The $30,000 includes travel for 2 people including airfare, hotel, cabs, meals and miscellaneous expenses. It covers booth rental for two booths, expenses on the show floor for labor, electricians, carpet cleaning and so forth. Drayage is a big cost in exhibiting and is the term used to describe the unloading of your crates, moving them to your spot, reloading them for storage and the reverse on the move out.

Other tradeshow expenses include printing and framing enough work to fill the booth, building crates to ship your goods and shipping your booth crates to the show. Advertising in the show directory and other promotional items are also considered in the figure. You can see the numbers add up quickly. Yes, you might get off with less than $30,000 to do a show, but it will probably come at the cost of eliminating something you wish you had done.

Can you start a self-publishing company for less? Absolutely, thousands of artists have. <u>The critical thing about financing is to build your expectations around your situation</u>. If you scale back on what you can put into the business, you have to scale back on what your expectations are on how much you will make from it and how long it will take you to get the reward and recognition that represents your long-term goals.

In order to raise funds and properly start your business, even if raising money is not an issue, you should know about break-even analysis. By demonstrating that you understand the concept and have developed realistic analysis for your business model, you will go a long way towards gaining confidence from investors and making them much more comfortable with you as a businessperson. The exercise will help give you confidence that you are embarking on a reality rather than a pipe dream. Demonstrating your knowledge of a break-even analysis will help you further impress investors, especially those who already believe in your art and your ability to continue to create new art.

Let's get into some of the pricing that goes on in the industry. To start, having a good understanding of how posters are priced through the various channels into the retail market will help you make better decisions about what you want to do with your art and will assist you in making good break-even analysis forecasts.

Typical Wholesale Poster Volume Pricing Scenario

A typical discount for a buyer to request from a publisher is for a 50-50-20 deal. You should be aware of this kind of industry idiosyncrasy discounting. Translated, the jargon means the buyer is asking the retail price to be brought down in price by less 50% less 50% less 20%.

Using a $40 retail poster price, let's examine how by honoring a 50-50-20-discount request a publisher's final wholesale net price is $8 per poster.

$40.00 x .**50** = $20.00 (First Discount @ 50%)

$20.00 x .**50** = $10.00 (Second Discount @ 50%)

$10.00 x .**20** = $8.00 (Third Discount @ 20%)

Don't be surprised when you find buyers who are not really volume buyers asking for similar discount. Many people in the industry don't think twice about trying to put publishers on their knees with regard to pricing. Some have legitimate reasons for asking, others are trying to bully you for a lowball price. Unfortunately, the above scenario is not the lowest price that publishers will be asked to provide for some buyers. And if the buyer is important enough to you, meaning they buy often and in quantity, you can expect to go to lower prices.

Keep in mind that the mom and pop galleries and frame shops that sell prints and posters are really small businesses that on average, according to *DECOR* magazine's regular reader surveys, generate less than $300,000 in annual sales, and that's average. Deduct rent, product, salaries and benefits, marketing and other costs all paid from that amount and you could quickly see why some have to be hardnosed on pricing. For them, it's the difference between profit and not.

There is a lesson here in that being sensitive to pricing and costs are things that should be important to your own business, so it should not be a shock that many of your smarter customers have the same concerns. How they come across when they ask for discounts is often the difference between wanting to work with someone or not. Unfortunately, like every other business, this one has its share of those who are difficult. The good news is they are more than offset by those who make doing business with them a pleasure.

What you need to do is figure out in advance what your bottom line is going to be for as many scenarios as you think you might encounter. There is nothing worse than being put on the spot with a tough question and a deal on the line and not being able to respond to it with authority. If you've done your homework, including your break-even analysis, and know what your costs are and what your inventory is, you can quickly calculate your reply to discounting requests. You'll even be farther ahead of the give and take if you have anticipated some requests and have a reply ready.

To keep things real, let me say here I don't believe many self-published artists are going to consider posters as their medium of choice. There are several reasons why.

First, the market has been changing for years with volume buyers becoming increasingly important to publishers. The number of independent shops continues to dwindle and those that remain are careful about what they stock. This gives new publishers fewer openings to be established with repeat buyers.

Second, selling posters these days requires having a great Rolodex full of potential volume buyers. Without important buyer contacts, a new poster publisher is faced with nearly insurmountable odds of getting to market.

Third, in order to be taken seriously by volume buyers and consolidators as Lieberman's, publishers need a full line of images. The days of starting with 8 or 10 images and building a line around them are gone for posters publishers.

The cost of getting into the poster publishing game has become prohibitively expensive while margins have become slimmer than ever. It's a business now where only the strong survive. Industry observers note that new poster publishers are not coming on to the scene any longer.

What is happening instead is a consolidation of publishers. Bentley Publishing Group alone has acquired Aaron Ashley, Rinehart Fine Arts and Leslie Levy Publishing in the past few years. The company also provides distribution for Grand Images. These are all publishers with large catalogs and impressive buyer lists. While their reasons aren't all the same for selling their businesses, all would agree that increased competitiveness and changing landscape helped them decide to sell.

Giclée Printing and Pricing

Given that serigraphy is too expensive for many new artists to produce, posters and open edition prints are a hard market to crack and offset limited editions have lost their luster, giclées have become the de facto medium of choice for self-published artists.

Giclée wholesale pricing is not typically commoditized and subjected to the intense discounting that you find in poster and open edition print pricing. A standard 50% off retail is the norm. That said, you will find pricing as you get into selling your prints at wholesale to vary widely.

A significant part of your pricing is based on your printing costs. The range of pricing from various ateliers for giclées can be confusing and confounding. You'll find there are numerous ways printers price their services, which makes it difficult to make compare them fairly. Not unlike trying to shop for a mortgage. Plain vanilla to plain vanilla comparisons don't exist.

The same situation arises with wholesaling giclées. While the standard 50% discount holds up better than in the poster market, you will be hit with galleries that will only take your art on consignment and still want 50% or more. This is a usual request from gallerists to make of an artist, especially an unproven artist. Be prepared to have to deal with requests for anywhere from 60/40 discounting in your favor to 40/60 going against you.

Bottom line is that you must do your spadework to get suitable answers to the questions you have about printing and selling wholesale. That means you need to dig for information by talking with lots of qualified people to before you will begin to feel comfortable with your choices regarding what vendors to use to print your images and how to deal with galleries when it comes to getting representation in them.

If you need help deciding how to price your work, consider contacting Aletta de Wal at Artist Career Training. She has built a formidable program designed to assist artists with this crucial component of their business. It is no overstatement to she is a guru of pricing for fine artists.

It would be great to offer you pat solutions to finding the best and most suitable giclée printer and for wholesale pricing to galleries, but there aren't any. Printers, for better or worse are a dime a dozen these days with more regularly joining the fray. You'll find small operations with a few people to large ones employing dozens. You have to investigate them, visit them personally if possible to meet their staff, see their quality and learn how they work with artists.

Here again, asking questions is a key weapon in your arsenal. Find out what printers the artists you admire are using. Read the trade mags to see what new services are being announced and advertised. Many topnotch giclée ateliers are not big advertisers or promoters, so you have to find them by following your nose. Since the creation of your image in printed form is as important as any one factor in your business, it behooves you to be diligent in working to find the one who will do the best job for you.

It was suggested earlier and is worth repeating here to either work with a member of the Giclée Publishers Association, or at a minimum work with a printer who adheres to their guidelines. Even that is not an ironclad guarantee you are going to be happy with your results. But, it's a good place to start.

Printers are the first part of the pricing and final cost for publishers. Other considerations include substrates, framing and fulfillment.

- Substrates are straightforward. Are you using canvas or paper or both?

- How will these choices affect the way you build and market your editions?

- Framing often is left to the artist, or the charges for framing are taken out before the split with galleries.

- How a piece is framed is a monetary concern to the artist, but a quality concern to the collector and to the gallery. Having worked in a gallery, my advice is to choose sturdy framing that can survive shipping and constant handling in the gallery. The gallery is responsible for damages after they receive the work, but you are better off sending them pieces in frames that are not fragile. You can count on your pieces being moved around a lot in the gallery.

- Fulfillment means that an important component of your business is warehousing and shipping your finished images. Some ateliers offer full service to their artists. They print the giclée, frame it, retouch or embellish on request of artist and ship the final product to the gallery or collector – all at a cost, of course.

For you, it's a matter of balancing what you are capable of doing, what you can afford and what is important.

A final consideration is that the cost of giclée printers continues to drop. That means for those artists who are techie or have staff that are, considering printing their own pieces becomes financially attractive. I urge great caution here, as I believe that it is more difficult than it seems, more time-consuming than it seems and has hidden costs to boot.

Worse, you buy a wonderful refurbished machine only to find that new processes have become available and that if you had waited, you could have had a brand new machine that does something totally cool for the same or less money, but you are stuck with payments on an nice adequate printer and a sunk feeling.

Because posters tend to be produced in traditional sizes, their prices are easy to predict. Even with the addition of mini-prints and oversized prints to the product mix, prices are within a range. Giclées on the other hand are not that predictable when it comes to pricing. Many aspects go into the pricing equation. They include: the artist's reputation; printing costs including substrate choices and finishing; size; retouching or remarquing; edition size; framing and fulfillment to name the more important components of pricing for giclées.

Regarding dealing with galleries, it's the same as with printers. You have do your homework and due diligent investigation before you proceed. In order to get your work in front of the decision maker and find out what they are willing to do for you, you must first prepare to find them and have your pitch and program pulled together and ready to present. A huge part of being having a program that is "pulled together" is having your pricing and all the elements that go into pricing accurately figured.

You will find, the more you have marketed your work in the trade publications and at tradeshows and through direct mail to them, the easier it will be to work out the most favorable deal possible with galleries and dealers.

It's all a learning process for you. The quicker you master it, the faster your success quotient will grow.

The figures for giclée pricing are arbitrary often sellers and collectors don't have easy apples-to-apples comparison.

To come up with the right conclusion to help you grow your business quickly, you have to thoroughly think things through on your selection of a printing process and your pricing plan. The outcome of your decisions will dictate your marketing and sales plan, too. All these items are interrelated and require that you use your best and most informed judgment among the various options open to you.

Since I am mentioning pricing requests and dealing with printers and galleries, what I am really talking about is an important skill that you probably haven't given much thought...the art of negotiation. If you or your marketing person can master or improve your negotiation skills, it will quicken your ability to gain a foothold in the market and grow your business.

Negotiation is a learned skill, as most of us are not instinctively born with the techniques. Sure, we all have learned how to bargain with our parents as children and later with our children as parents and our spouses, too. In business, however, it is different, more difficult and formal. Once you have a grasp of the basic techniques of negotiating, you are going to close more deals that are beneficial to you than if you are winging it and learning how to negotiate with OJT (On the Job Training).

One of the better books on the subject is *Guerilla Negotiating: Unconventional Weapons and Tactics to Get What You Want* by Conrad Levinson, Mark S. A. Smith and Orvel Ray Wilson. It covers the topic with easy-to-understand concepts that explain how to get a "fair" advantage from a negotiation. Another terrific book on the subject is *Getting Past No: Negotiating Your Way from Confrontation to Cooperation* by William Ury. Sadly, learning to effectively negotiate is not likely to have been included in any of the courses you may have taken in your life with regard to art or even the business of art. **Please use this "head's up" on importance of negotiating and begin learning about it now!**

When you think about it all artists, even those who are museum bound fine artists at the top levels who will never publish a single print, are in constant negotiation throughout their careers. Whether it's with agents, galleries, collectors, charities, landlords or others who want something from the artist, it always comes down to negotiation.

Take charge of this for your professional life and it can make a major difference in the outcome of your success. I urge you to spend the time to read and study on the subject of negotiating and put into practice what you learn. Over the course of your career, the time spent to learn how to better negotiate could become some of the most valuable you ever invest. It's your career, your future. Why put yourself at a significant disadvantage in this crucial area that can affect every aspect of your career?

Will Your Self-publishing Art Print Business Make Money?

You won't have a chance of knowing that without doing some calculations and one absolute must is to prepare a break-even analysis before doing the more detailed work on your overall business plan.

While you can't know with utter certainty that your business is going to be profitable, you can get much closer to a realistic idea by analyzing the financial accuracy of your concept. Perhaps the most basic research you can do is to prepare a "break-even analysis" for your business. This, along with a few other financial forecasts, will give you and potential investors a more clear determination of whether your business will succeed.

A break-even analysis shows how much revenue you're business will have to generate to cover its expenses before profits. Your obvious goal is to promptly surpass your break-even point so you are not only covering expenses, but generating what is called cash flow, or what's left after expenses. You quickly need to bring in more than the amount of sales revenue you need to meet your expenses. By doing this, you significantly strengthen your odds of seeing your business prosper.

You will find that both well-informed investors and entrepreneurs employ a break-even analysis as their main tool to sift through ideas for new businesses. They look for projected sales revenue that surpasses their costs of doing business because they know it's crucial to get this kind of financial justification from their break-even forecast before taking the next step of writing a full business plan. Since having a break-even analysis will be an integral part of your business plan, you give yourself a head start on getting that important piece of pre-launch planning done.

Can you skip on this work? You bet you can, but "bet" it becomes the operative word to describe it because it is an enormous gamble to launch a business without proper financial preparation. Remember, sophisticated investors won't allow you to forgo doing this important prep work and not having it prepared will negatively influence their impression of you by showing them a significant lack of business skills and attention to details. You can fly by the seat of your pants with your own money, but you won't get off the ground that way with investors. If this is sounding like too much work, not fun or in some other way not worth doing, it's a sign that you should be seek a publisher instead of self-publishing.

Here Are the Steps to Prepare a Break-Even Analysis

To start with, you'll have to use a system better than the S.W.A.G Factor I mentioned in Chapter Five to hone in on what your expected expenses and revenues will be. This will require some valuable intense research from you. It should include a competitive market analysis, which is an investigation of companies in the market place and in particular, what their pricing strategies are as best as you can ascertain.

Unfortunately, since the art business is nearly 100% privately owned, you won't find much in the way of published sales figures and revenues to base your own projections on. To a great degree, you'll have to wing it on this one. But by doggedly asking questions and getting help from the right individuals, you can come up with a realistic projection on the business.

Needing facts, it will serve you well to befriend knowledgeable people in the business. Other artists who are already enjoying success, publishers, editors, sales reps or anybody who can give you information that will help you formulate a good overview of your competitive market should be cultivated by you and your organization, particularly while it is in the budding stage.

A good source rarely tapped is the moulding rep. They know who is buying the quantities and often have insights into the businesses that are their customers. If they can't tell you specifics, they ought to be able to generalize in ways that will still be helpful to you. Find out whom reps Larson-Juhl, Roma, LaMarche or some of the other top moulding companies in your area and contact them. Reverse the tables and treat them to lunch. They can be an invaluable source of all kinds of industry news and insight for you.

You don't have to know details about all the businesses, but it helps to know as much as possible about those whom you see as your main competitors. You will find that knowing your competitors isn't just a way to spread industry gossip; it will help you make critical decisions that can help your company grow.

Getting a copy of any of the latest research by industry publications like *DECOR*, *Art Business News* or *Art World News* would be about the best source you will find to help you begin your market analysis. The Internet remains a great source of information and using all available industry contacts can be useful, too.

You'll also need to determine your own projected sales volume and your anticipated expenses. It's a worthy idea to spend some time and money on books about business planning. Some excellent software also is available to teach you how to make reasonable revenue and cost estimates. And, you can find remarkably good free advice and help on this and other startup questions through SCORE (Service Corps of Retired Executives), which is a non-profit organization that is part of the U.S. government Small Business Association. You can find your local chapter online at: *www.score.org*

Break-even analysis is critical to a start-up business. You will need the following figures and calculations to complete your analysis:

Fixed Costs – These are also called overhead. These figures don't vary much from month to month. They include rent, insurance, utilities, and other set expenses. You also should put in a line for miscellaneous costs in the amount of about 10% of the total of your other fixed costs to help cover expenses you can't predict.

Sales Revenue – Total all the money your business will generate in sales each month or year. Please don't let your enthusiasm or desire to make your numbers work out influence you and make you avoid being brutally honest in evaluating your sales potential. It will come back to haunt you like a bad dream. You have to base your forecast on the volume of business you can truly expect and not on how much you need to make a good profit.

Average Gross Profit - This is what is left from your sales revenue after you pay the direct costs of a sale. (Direct costs are what you pay to provide your product or service.) Here's an example: If your giclée print costs $100 to produce and you wholesale to a dealer for an average of $300, your average gross profit is $200.

Average Gross Profit Percentage - This calculation works off the figures you pulled together for your Average Gross Profit Per Sale. It shows how much of each dollar of sales income is gross profit. To get the answer, divide your average gross profit figure by your average selling price. From the above example, take your average gross profit of $200 for your giclée' print and divide by the average of sale price of $300 and you'll show a gross profit percentage of 66.7%.

Calculating Your Break-Even Point – Now that you've done the above calculations, it's simple to figure out your break-even point. Divide your estimated monthly fixed costs by your monthly gross profit percentage to learn the amount of sales revenue you have to generate to break even.

Staying with the example of the wholesale giclée business, let's presume your fixed costs average $4,000 per month. Now applying your anticipated profit margin of 66.7%, your break-even point is $6,000 in sales revenue per month ($4,000 divided by .667). Therefore, you must have monthly sales of $6,000 to cover your fixed costs and direct (product) costs. If you want to eat and grow your business, you have to make more than that because that $6,000 doesn't include a profit for your company or a salary for you.

You wouldn't be the first brave entrepreneur to go through this financial exercise only to be sadly surprised that you didn't hit a break-even point. If you find it is higher than your anticipated revenues, you will have to go back to the drawing board to see if you can change parts of your plan to reach a break-even point.

Here are some things you might be able to do to cut your costs:

- Charge more for your work
- Find a less expensive supplier or buy in larger quantities to get your costs down
- Cut down on personnel costs, maybe you don't need someone full-time right away

- Work from your home or share space somewhere instead of paying rent
- Use your creativity to figure other ways to cut down overhead

Cutting costs should be part of your business, at the core part of your business fiber, regardless of how successful you are. You'll find few if any top executives at any company, art or otherwise, who aren't extremely sensitive about keeping costs in line. As the wise saying from Ben Franklin goes, "A penny saved is a penny earned!" If you start with this as an integral part of your business plan and actions, you'll give yourself greater chances for success early and better profits later.

If you have massaged the numbers and your break-even sales revenue still isn't working out satisfactorily, it might be time to back away from your business plan. If that's what happens, it's not the worse thing. Imagine if you didn't do the calculations and ended up blowing a lot of your money and maybe even some of your friends and family who didn't require hardheaded planning and forecasting before passing the hat for you. Remember what I said earlier, a viable option for some of you is to consider taking the vacation of a lifetime — and if so, Bora Bora awaits you.

Success takes more than a good break-even analysis. If your forecast shows you'll generate revenue beyond your break-even point, you are fortunate and moving in the right direction. But that only means you now will need to figure out how much profit your business will generate, and whether you'll be able to pay your bills on time when they are due and more. Your positive break-even forecast is a great start, but you will need a more complete analysis before moving from hypothetical to actual in putting dollars into your art business.

Here are some other necessary projections to complete your financial picture. Since they also will be an integral part of your business plan, they are especially worth completing:

- Profit-And-Loss Forecast – a month-to-month projection of net profits from your business operations;
- Cash Flow Projection – shows how much actual cash you'll generate each month to meet your expenses;
- Start-Up Costs Estimate – shows the total of all of your expenses incurred before your business opens.

There are many resources for finding out more about completing a business plan and conducting the exercises to get accurate financial projections for your business. Here are some suggested titles:

The Entrepreneur's Success Kit: A 5-Step Lesson Plan to Create and Grow Your Own Business by Kaleil Isaza Tuzman

The Definitive Business Plan: The Fast Track to Intelligent Business Planning for Executives and Entrepreneurs (2nd Edition) by Richard Stutely

Small Business: An Entrepreneur's Business Plan by J. D. Ryan, Gail Hiduke

A great suggestion is to get the bestselling business planning software: *Palo Alto Business Plan Pro.* The software receives high marks from numerous reviewers. I've used it myself and found it to do all expected and more.

> I went into the business for the money, and the art grew out of it. If people are disillusioned by that remark, I can't help it. It's the truth.
>
> – Charlie Chaplin

Chapter Six

Marketing and Selling Your Self-published Images

> **Art is making something out of nothing and selling it.**
>
> – Frank Zappa

Those wishing to learn more about the art business are blessed with an abundance of books on the subject. In my research for this book, I was amazed to find a lack of information in them on the important topic of art print publishing. This was odd to me because from my observations over the years getting into the print market made a huge difference in the careers of artists.

The realization existed gave me impetus to finish this book and get it to you.

As with any business venture, there are always basic points that can be considered either merits or pitfalls. Here are some for you to reflect on:

Self-publishing Merits

- Artistic Control
- Make More Money
- Produce at your own rate

Self-publishing Pitfalls

- Need help, as in employees

- Need financing
- Your art, i.e., your passion, is a business
- Odds of success are determined by your desire and time to be involved in most, if not all aspects, of the business to make the venture successful, especially at the outset

To begin the process of becoming a self-published artist you must first have a good understanding of your target market. If that piece of advice sounds elementary, it is. Nevertheless, it is crucially important for you to know and understand your target market — to have in mind the person who will be buying your prints.

Best Buy recently did a study of its customers and realized that 20% of them were responsible for 80% of its business. It also found that many of its customers added little to the bottom line. In fact, another 20% were the cause of much wasted employee time, consumed much of the company's promotional budget and were the least profitable in terms of sales. The company refocused its advertising and retrained the employees to cater to that precious 20% of the market and saw sales improve by more than 20% in a matter of months. While you might not have the resources of Best Buy, you can still take a page from its book and work at learning to identify your target customers. When you know who they are, you can figure out how to cater to them.

If you are the Snap-On Tool Company, you know your primary target market is men ages 21-55. You manufacture tools to appeal to them and likewise your marketing. It's not to say that art and mechanics tools have a lot in common although I do find a sculptural value to many well-made tools. (The woodworker in me comes out again.) To reiterate what I emphasized earlier in this book — don't let go of the idea that even though it's art and it is beautifully created with loving passionate care, it's still a business if you want to earn a living from it. I believe the way you market your business easily can make a 20% difference in your results, regardless of your size.

You know what the subject matter of your art is and you probably already have a good feel for who are the dominant buyers. If you don't know this information by research, experience and instincts, you need to figure it out because it is vital to have a clear-cut vision of the niche you are going after. Who are you painting for; who is your buyer? If you can't describe your target audience, finding buyers is going to be difficult.

The more clearly you can see your prospective collectors, the more likely you are to create imagery that will appeal to them. Is thinking about marketing your art a way of selling your artistic soul? It doesn't have to be. It depends on how you conduct your business. Finding a comfortable way to feed your artistic body does not require selling your artistic soul.

To ensure your success, you should strive to become a student of your collector base. Start to study them and when you begin to understand why people might buy your art and what is it about your art that appeals to them, you are getting closer to honing your success. Don't take the S.W.A.G. Factor approach by doing your research anecdotally; use a facts-based system and accurate measuring. Be aware your mind can play tricks on you because it will default to your own prejudices if you don't track the information about your buyers as you receive it.

Find a way to keep score of not only who buys, but also who shows an interest and who doesn't. While you are sleuthing away, also track what it is that is selling and attracting interest. You will see patterns evolving from this activity. It will tell you what images in your line are working. Continue to refine and use the information to determine if it is content, size, color, price or some combination of those that is generating sales and interest.

Make it a point to ask people questions, which pieces they like and why. Be honest and ask for their help. Tell them you are doing your own informal research. Sometimes in a direct sales situation as at an art show, approaching them in a non-sale, non-threatening manner gives you an opportunity to talk about your art without showing you are eager to make a sale to them. This approach coincidentally will lead to sales when prospects relax with you in this process. It could be as simple as this, "You seem interested in that piece. I'm doing some informal research on my work to find out what it is about certain images that appeal to people. Would you mind sharing with me what drew you to the piece?"

The purpose of doing personal market research is to begin to learn which of your original paintings you should consider turning into fine-art prints. While you must know what to paint for your audience, you must also be able identify what niche you are aiming at with your art. With the data you gather, you will make informed determinations about what your target audience expects in pricing, as well as content, color and so forth. Gaining a clearer picture about these things will provide the support for your decision with regard to which is the best reproductive process for you to pursue, i.e., poster, limited edition offset prints, giclées and serigraphs.

So, What Business Are You In?

Many businesses don't know how to correctly answer that question and they pay a hefty price for it. Ray Kroc became a billionaire with McDonald's because he realized he was in the real estate leasing business. If he had thought he was in the hamburger or restaurant business, he would have made many errors in judgment on how to build his business model. Likewise, if you consider yourself in the "Art" business, or even the "Art Publishing" business, you are off on the wrong foot and headed up the wrong path.

Okay, what business are you in? **You are in the business of building, nurturing and replenishing a dealer network.** It is through that network your artwork will flow to the end consumer. It is business of the art retailer to romance your work to "their" customers. Their customers then become your collectors via your wholesale network. The primary job for your publishing business is to find and romance art dealers.

Think of it this way:

- Build
- Nourish
- Replenish

Put another way, it means Job One for your business is to come to work everyday with the goal of building that distribution channel of art retailers and art dealers. If you are a poster publishing company, you also will be focusing on the OEM or volume framer market.

OEMs & Volume Framers

A major part of the distribution system to the mass market for posters and open edition prints requires an understanding how the volume framer side of the business affects the market. Volume framers are sometimes called OEMs.

The acronym OEM stands for Original Equipment Manufacturer. It denotes a company that uses parts jobbed out from other manufacturers and assembles them under its name or the trade name of another company with whom it has contracts for assembly work.

Frankly, OEM is a more common moniker for those in heavy manufacturing than the art business, but all the same, when you see art in a matted finished frame at Pier One, Crate and Barrel, a furniture store or any other mass retailer, it almost always will have come from an OEM or volume framer. OEMs buy prints and posters here, matboard and glass there, picture frame-molding elsewhere and do the final assembly and marketing of the finished product.

These volume framer companies are not easy to identify because they don't market to consumers and have names that are unfamiliar outside the industry. A great suggestion for finding many of them is to search through the directories of companies that exhibit at the IHFC (International Home Furnishings Center) in High Point, North Carolina. If you haven't heard of High Point, you need to learn about it because what happens there dictates home furnishings fashion trends, including wall art.

The Web address for the IFHC is: *www.ihfc.com.* Search it for companies that are wall art vendors. These companies are either volume framers or jobbing out images, matboard, glass and picture frame molding specs to mass or volume framers to produce for them. Having a buyer from one of these companies can be a huge break for a new publisher. Unfortunately, getting to them is not that easy. Your best bet is to establish a complete marketing and promotional campaign that includes getting you on their radar in due time.

Twice a year in spring and fall, the IHFC runs for 11 days and brings in more than 100,000 buyers from all over the world to see the offerings of companies selling furniture and decorative accessories to the home furnishing retailers. There are other organizations, displaying more home-furnishing vendors in multi-story buildings, who compete with the IHFC facilities. Some are temporary and others are permanent. Some of those facilities include: Market Square, Hamilton Market, and National Furniture Mart.

The total showroom space in and around High Point is estimated to be over 10 million square feet. Its size is astounding and overwhelming for any buyer who would attempt to see it all. As a way of comparison, the Decor Expo Atlanta show, which at nearly 2,000 booths and is itself huge (it's listed as one of the 200 largest shows in the country) requires about 360,000 square feet for its show needs. That means you could fit nearly 30 Decor Expo Atlanta shows in High Point. If you have ever walked that show, you know how astonishing this comparison is.

What's does this size mean? Literally, the appetite for artwork is vast and these companies represent the potential for sales in the thousand of units. Once you are established with them, you are in the gravy. Are any of these companies likely to do business with a rookie self-published artist? While the short answer is no, it doesn't mean they shouldn't be part of a longer-range plan for you.

The importance of this market segment is underscored by the fact Pfingsten Publishing LLC launched in 2004 a new publication called *Volume* to address this market. Pfingsten also is the publisher and producer of *DECOR*, *Art Business News*, ArtExpo, Decor Expo and other industry publications. The following market description and information is taken directly from the *Volume* advertising media kit:

The United States retail custom framing market has approximately $4 billion in revenues and has recorded an annual growth rate of approximately 6 percent over the past five years. The retail custom framing market is highly fragmented, with independent, non-franchised stores accounting for approximately 80 percent of the 17,000 framing stores in the United States and 89 percent ($3.6 billion) of the revenues. Retail chains and their franchisees account for the remaining 20 percent of framing stores in the United States with 11 percent ($400 million) of the revenues.

Over the last five years, the retail distribution channel for wall art décor — mass-produced, framed ready-to-hang artwork — has grown exponentially. Large national and regional retailers have taken advantage of economies of scale, and marketing leverage, to put pressure on the manufacturing sector of the industry. The result is the production of high quality pre-framed art at price points substantially lower than that of custom-framed art. *Volume* speaks to the manufacturers that have changed with the industry, and have developed high-volume art and framing outlets that cater to national retailers.

Audience

Volume is a controlled circulation publication, provided to qualified subscribers free of charge. However, readers will be required to complete a request card providing detailed information on their business. Subscriptions to

Volume will be based on meeting a certain minimum demographic requirement. This requirement is necessary to ensure that *Volume's* circulation is highly targeted, and that the magazine reaches key decision-makers in the market, while at the same time, maintaining high integrity with the advertisers that support this market segment. *Volume* will premier with a circulation of roughly 3,750 volume framers and OEMs created by combining lists from major companies in the industry.

Editorial

Volume's mission, as with Pfingsten's other art & framing markets, is simple: to bring together contract and volume framers with molding, equipment and supply manufacturers and distributors to help each do more business. Given this unparalleled mission, the editorial content in *Volume* is unlike any other industry publication. *Volume's* editorial content is focused on the specific needs of the contract framer, from economic concerns to manufacturing techniques.

Launching new publications is at least as chancy as starting an art print publishing venture, perhaps more due to startup costs. Even a company with Pfingsten's experience and financial backing will have to work hard and smart to make *Volume* profitable. Don't' be surprised if you are reading this book sometime after it's initial publication in March 2005 to find Volume had the plug pulled on it. I wish the publisher well and appreciate the concept behind the book. Still, I have to be realistic in assessing their chances, which I rate dicey, which describes the statistics of publishing startups in general.

Okay, back to the burning question: If building a dealer network is my business, how do I do it? The answer is simple and easier to say than do...<u>one at a time</u>. There arguably are somewhere between 13,000 – 20,000 art retailers in the U.S. that sell reproductions of some sort, as opposed to fine art galleries that exclusively or primarily deal in originals. If you annually could add 1% of the low side of the range, gaining 130 new dealers a year, you'd be doing a smash up job.

Let's look at this scenario. Suppose you were in a limited edition business as giclées and you had a dealer base of 130. Make a further supposition each dealer averaged one sale a month for you. That means you'd be shipping more than 1,500 prints to them annually. Based on a conservative $500 retail price point, you would net $390,000 in a year before any expenses. Of course, that is not your break-even figure. But, if you deduct printing costs of $156,000, it leaves you $234,000, which would more than cover the $119,000, (which budgeted $12,000 for printing), arbitrary figure I provided earlier as an example of what the possible annual cost might be to run a publishing company.

The admonishment, "Your results may vary," is apropos here. These are example figures and every situation is unique. Your pricing could be much higher or lower than $500. For most, gaining 65 new dealers in the first year would be a major accomplishment. However, if you sold them each two prints per month, you would have the same total sales as the original example above. So many variables come into play it makes it difficult to know in advance what your results will be.

Setting the popularity of your art aside, your results are going to depend on financing, marketing execution and follow up; how in tune your images are with market trends (if you are leading its great, lagging not so good, even being too far ahead can be hurtful).

> If you're riding' ahead of the herd, take a look back every now and then to make sure it's still there.
>
> — Will Rogers

Sometimes things as the dreadful 9/11 attacks knock the sails out of the best-laid plans. Ask any of the hundreds of companies who were devastated by that event. Many were looking forward to one of the best shows ever in Atlanta (scheduled to begin September 14, 2001) only to see it cancelled in the wake of the attacks on that day of infamy.

By early 2005, many of those companies had not fully recovered from the losses of that show's cancellation or from the repercussions that rippled the economy in its path. You see, for many of the companies that exhibit in Atlanta, it is Christmas for retailers. They rely on it for a large part of their annual sales and for the invaluable contacts they generate there each year. If you get in the art print game, you will too.

Now even though you have some visions of sugar plums dancing in your head as you contemplate these financial scenarios, you need to keep in mind that newcomers always face a tough, competitive rough road filled with many obstacles before it gets smoothed out. Yet, for the bold, undaunted and well prepared, the rewards can easily be worth the effort. I will always contend that no matter how crowded a competitive field gets, those with talent, grit, smarts, financing and ambition will not only find a way to survive, but will thrive.

Let's get back to building a dealer network. Yes, it is done one at a time. It takes a combination of coordinated marketing components to effectively move the sales dial in your favor. You and your marketing maven partner need to gain a good understanding of how these various components work and learn how to blend the use of them into a seamless marketing strategy for your art business.

The more planning you put into the details, the less likely you will be managing by crisis. Even though this all might sound overwhelming, I assure you being in the position of constantly being late or missing marketing opportunities is even more overwhelming. Fortunately, there really are only a few ways to get to market. They include:

- Advertising
- Tradeshows
- Publicity & Promotion
- Direct Mail
- Email/Website
- Reps

Other items could be listed above. They were left out intentionally because they are so far less effective than those listed and aren't worth mentioning. Be thankful you don't have newspapers, television and radio to contend with, as do marketers in other fields.

Comments on the items listed follows below in hierarchical form with the most important first.

Advertising

It's true. I sold advertising for *DECOR* magazine for 15 years, which naturally predisposes me to believe it is the most important. But, I also sold tradeshow space, direct mail lists, Websites and more. I learned from first hand experience with all methods what worked for my customers.

Moreover, every bit of research I have ever seen from any trade publication serving the retail art industry conducted with their readers indicates that magazine advertising is the most influential of all the ways dealers use to find new companies and new images. With their monthly reach, magazines are the most important sources of information for buyers.

Magazines tend to be saved and referred to so they have shelf life for your marketing message. Keep in mind, I no longer sell for *DECOR* and haven't for several years, which means I gain no advantage by making these statements. I say it here because I deem it is the best advice I can give you.

When you ask what you should do with regard to advertising versus tradeshows, my answer is to do as much of each, along with the other items mentioned above, as humanly and financially possible. Doing so is the best way to get the maximum exposure in the least amount of time. Mass marketers measure their impressions in the millions and know that getting enough exposure with the right message will move product. You have the same task on a smaller scale. You want to make as many impressions from as many angles as possible on your target audience.

When you begin to plan your advertising, start with the idea of using it for all its purposes. If you think of it as a vehicle to merely show your latest work, you are shortchanging yourself and your advertising's potential. To maximize return on your investment in advertising, use it to coalesce your message being streamed to your customers and prospects in other ways.

In planning, use the four Cs of advertising:

1. Clarity

2.Coordination

3.Consistency

4.Continuity

Clarity – Don't complicate your message. Be clear about who you are and who you are not. You don't have to cram every aspect of your business plan in each message. Be concise and clear and keep it simple.

Coordination - Use your advertising to coordinate the message, look and feel of your entire marketing efforts including, logos, direct mail, Websites, email, et cetera. For example, ads for Coca-Cola may be strikingly different from each other, but there is never a doubt as to whose ads they are. The same is true if you get coupons, postcards, or emails promoting Coke. Maintain a coordinated appearance in the integrity of and look of your company image and brand as they do.

Consistency – Decide what your message will be and maintain a consistency with it in all of your advertising. The impact of your advertising becomes exponential when it is consistent over time. Your customers and prospects will not only see your latest offerings, but be reinforced with your message.

"Nobody doesn't like Sara Lee"

"Budweiser, The King of Beers"

"Burger King, Have It Your Way"

Those tag lines or slogans don't necessarily dominate every ad from those brands, but they have appeared often enough that they are instantly recognizable because they have been consistently present in message after message from their marketers.

By maintaining a consistency in your advertising, you lay a foundation and build upon it. The ones that come before support each succeeding ad. By doing this, you strengthen your message as it is consistently reinforced to your customers and prospects.

Continuity – When millions of people can sign your jingle or repeat your advertising slogan, isn't the battle won? Why would McDonald's continue to advertise when they have enormous visibility in great retail locations, millions of customers coming into their restaurants daily and so on? It's not because they want you to enjoy the free programming you see on your television; they are not that altruistic. It is because they know their message needs to continue to be repeated in consistent manner in order for them to continue to capture market share. The adage, "Out of sight, out of mind" applies to mass marketers as much as it does to growing publishing companies and hopeful artists.

Deciding to run one or two ads to see what happens before you commit to more is throwing your money away. Your prospects in trade magazines are retailers who are see new artists and publishing companies coming into the marketplace all of the time. They are waiting for more than two ads to see what happens to you.

What dealers want to see is more than an ad with beautiful artwork displayed. They want to see commitment from the advertiser that they believe in their work and their message. The last thing they want to do is commit their time and effort to bring in a new artist, sell a few pieces and find out the artist has abandoned the print market. One of the best ways to convey your conviction alongside your fabulous images is to use continuity or constancy in your advertising.

Research over the years in various media has shown it takes as many as six exposures to a message before someone will react to it and respond. This is why I put a minimum of six ads in the proposed budget for a new advertiser. If finances are thin, you can of course do less, but you will have to realize your effectiveness is going to be diminished. If this is the case, you should scale back your expectations to fit your budget. The other thing to do is to work harder at publicity and promotion to help balance out a reduced schedule.

Years ago, the market between open edition and the higher priced limited edition prints was split by the marketers themselves between *DECOR* for posters and *Art Business News* for higher priced work. That played out as well with the tradeshows where the former attended Decor Expo and the latter ArtExpo.

With the merger of these properties into the Pfingsten Publishing's International Art and Framing Group, those obvious distinctions have changed some. You will need to keep up with who is doing what and where to make sure you are advertising in the proper vehicle. Study the trade publications for more than the obvious. Learning to be an astute observer and to glean information from them will help you hone your competitive advantages.

When *Art World News* came on the scene, it immediately set out to compete with *Art Business News* and has done an effective job of it since. You'll notice some advertisers only advertise in *Art World News.* That's interesting since there are no priority points earned from their advertising to be used towards placement in tradeshows and so forth for advertising in AWN. That should tell you something about the efficacy of *Art World News.* If you intend to publish higher priced prints, you should have a conversation with the staff at AWN to ask why they get and keep those exclusive ad contracts.

Perhaps you could even ask some of those exclusive advertisers in AWN why they made their decision. If you are sensing you need to act in your own best interest and make your decisions based on the best information you can garner regarding your marketing, you are right.

There is always jockeying among advertisers, particularly frequent advertisers who demand that their insertion in the magazine be placed in a FFRHP (far forward, right hand page) position. You'll notice in many publications, some advertisers have franchised a certain page upon which they appear every month. They normally get this placement by agreeing to advertise every month and paying a premium on top of their regular ad rate to retain that position

My opinion on premium positions has always been that you can negotiate for some decent placement without paying a premium. While you may not get a specified page, you might be able to swing a right hand page in the first 50-70 pages, by agreeing to a multi-issue contract. That way the money saved on premium position can be used instead to buy more space.

For example, if you are saving a 10% premium that means you could run 11 ads for the cost of 10 with 10% premium — give me the 11 ads every time over position. Of course, the longer your contract the better your negotiating position, but a six-month contract ought to be enough to get you some consideration. Again, your negotiation skills will come in handy here.

Sometimes, pages that are adjacent to some of the best-read copy in the publication go begging with little or no demand for them. This situation happens because often there are departments, table of contents, masthead, columns and clusters of competing facing full page ads all vying for upfront positions even though the feature copy that is usually most important to readers is farther back in the publication. And, unlike aggressive competitive advertisers, readers aren't keeping score of whose ad appears where in the magazine.

A well-executed ad will pull better response than one slapped together regardless of where either is placed in the magazine. There have been tests to show that placement is overrated, and I agree. I still prefer a right hand page facing a full page of editorial in the first feature section of any magazine in which I was advertising. Since the fiercest competition is for the front and cover spots, you can often land this kind of position by asking (negotiating) for it.

A word about well-executed ads; they don't just happen. There is rhyme and reason behind how well some ads pull over others. It was always frustrating for me to work with artists who designed their own ads and had no clue what they were doing. Being blessed with artistic skills doesn't mean you naturally have great graphic art skills.

The advent of the PC and the Mac, along with the arrival of Illustrator, Photoshop and other desktop publishing software, gave everyone who could work those programs the idea they were a graphic designer...wrong. Even if you have the sensibility of an artist, unless you can couple that with the training of a graphic artist, you are probably <u>not</u> going to be able to do the best job designing your own ad. Besides, you are probably too close to the subject matter to be objective about it anyway.

There is more to putting together a good ad than including a graphic of some beautiful image that you painted. Just being able to understand typography and type placement is huge. Knowing how certain colors reproduce in magazines can be huge. Knowing why certain color combinations work better in some ads is helpful.

Knowing how to properly create a proof can be huge. If you, for instance, proof your self-made ad on beautiful snow-white proof paper, it's not going to appear that way on the typical grayish off-white 70 lb coated stock magazines use. You should always pull your proofs on magazine based stock. Otherwise, you aren't comparing apples to apples and you are setup for certain disappointment in your ad's reproduction in the publication.

And, the list goes on from there in lost opportunity to put your best foot forward by doing it yourself. Do yourself a favor, hire the best graphic designer you can afford and have them work with you to pull together your whole graphic style, not just for ads, but for logos and all the other elements that make up awareness for your company.

Let's review, you have agreed to pay $3,000 per ad to get into a publication, but to save some money, you decide to put the ad together yourself using limited or no background in graphic arts. I've seen this scenario often and it yields results that are less than satisfactory. You don't get the best response you could have from your ad and you don't stand out in a field crowded with competitors seeking to sell to your customers and keep you from getting theirs.

To reiterate this often overlooked point; if you don't have training in graphic arts, find someone who does, <u>and check their resume and portfolio before you do and make sure that the portfolio is really theirs.</u> I wouldn't compound the situation by hiring someone who is self-trained unless they had extensive experience, a great sample book and references. I don't have anything against self-trained people, many are capable, but in a world where there are many qualified and educated people to do the job, why take the chance?

Some advertisers look upon magazine advertising as a beauty contest, especially in the art business. It's not limited to art magazines. You still see high-powered Madison Avenue agencies concocting clever memorable television ads that do everything but help the viewer recall their advertiser and what they wanted the viewer to do. Your ad should not just display your art, it should convey a sense of your company and sell something, or ask for the order.

These folks running beauty contest ads are not properly answering the question, "What business am I in?" They think they are in the art business when they should be concentrating instead on building their dealer network. You should be happy that they are helping you out by not taking full advantage of their advertising dollar.

David Ogilvy, who was a giant figure in the development of advertising in the 20th Century, taught every ad should sell. He championed the concept of quality in visual taste and literate copy in advertising. You could do no better than to follow this simple powerful bit of advice for your own advertising:

> Always give your product a first-class ticket through life.
>
> – David Ogilvy

I'm telling you this because through my own observations over nearly two decades that often the ads that pulled the highest number of reader responses weren't necessarily the prettiest. Many had multiple images in the same ad, (They were showing their product...gasp!) although most weren't neatly lined up like tombstones in a military graveyard. Rather, they would be overlapping at attractive angles or other interesting layouts. Don't be afraid to show the product. Do it as David Ogilvy advises, "first-class."

Trade publications in this field have done away with what were called "Bingo Cards." Those were the ever-present postcards with rows and columns of numbers on them. Readers were instructed to circle on the card the corresponding reader circle number found at the bottom of any ad they liked and return the card postage paid to the magazine to be compiled and forwarded to the advertiser.

Bingo cards were always faulty. Ask yourself, how many good prospects would use them instead of using the phone? And, how much value you would receive from the many great literature collectors who were the primary users of bingo cards? Nonetheless, they did give advertisers and publishers some independent means of measuring response. Now with toll-free phone lines, email, Web pages and even toll-free fax numbers, they have lost their usefulness and really only add unnecessary expense to producing a magazine, and the cards have now gone the way of the free meal on an airline.

You can do some of your own research by training your staff that answers the phone to ask the person calling where they saw your ad. If you are in several publications, you can run different ads in the same months to help avoid confusion. Many new phone systems have extensions that can be keyed specifically to an ad in a publication. You accomplish this, the better and more thorough you are at doing it, the better your results will be. Offer different specials or challenge your creativity to come up with other bright ways to key response to your ads.

The most important thing to do is to get a consistent reliable system in place to help you monitor your advertising so you can track where your response is coming from. David Ogilvy would also tell you if you continuously test your advertising, you would continuously get better results from it.

Some advertisers go to the extreme of entering the data in a prospect database to track what ads pulled in a new prospect, but also tell what ad pulled a response that ended in a sale. That's a lot of work and is beyond the scope for a new publishing business to go that far, but it is something to shoot for in time. Knowing how to best allocate your advertising budget gives you a tremendous advantage over less sophisticated competitors.

Magazines have to be more imaginative these days to come up with ways to help their advertisers get and measure response. One of your duties is to debrief your rep on everything the publication will do for you to help you in this regard.

I would ask my rep to let me know what ads pulled the best response in the last 12 months if they have any way of capturing information that can be helpful to you. It is not a sure thing you will get the research. If you do, it will help you understand what kinds of ads pull best from a graphics point of view, but also what is pulling well from a content, size and color point of view as well. Be curious and persistent in probing for information your vendors have to offer you. Knowledge is power.

Some publications have value added programs they offer to advertisers. These are not always thrown on the table; you should make a point to always ask what is available.

Will they do an email or fax broadcast for you? Will they give you use of their mailing list at no or reduced costs? Are there package deals you should know about? What tie-in editorial promotions do they have? Often, if you don't ask these programs won't be revealed since most of this stuff is a nuisance with little ROI (Return on Investment) attached to them for the publications.

Advertising is a mix of part art and part science. You want the finest looking, most creative and inspiring ad that will do the best job for you. You will have multiple purposes for your advertising, but "to sell" should always be part of the purpose. It will be your front line in helping you create a brand strategy for you, your art and your company.

If branding is your top priority, you might want to forgo the idea of placing a bunch of small ads that would try to maximize response in favor of something that is planned strategically to lift awareness for you and your images. The latter concept could be a follow up to a series of ads scheduled as branding — keep thinking long range. And, if you are branding, the top-notch graphics are that much more important.

When Greg Bloch launched Triad Art Group Publishing, *www.royoart.com*, to sell the works of the fabulous Spanish artist, Royo, he spent far more on his ad budget than one normally would if using a typical 5-12% of annual revenue figure. He knew that by running a slate of full-page and spread ads with Royo's dramatic, romantic images would leverage his brand and his artist and ramp up his sales at the same time. And, he knew that this aggressive advertising would make his company seem larger and more important than it might otherwise have been perceived, given that it was a startup.

Greg came into this situation prepared with experience, financial wherewithal and an artist ready to catch fire. His results were nothing short of phenomenal. His advertising allowed him to compete with larger, established publishers and set the stage for his marketing efforts that were heavily focused on selling pre-show and at-show at ArtExpo.

Certainly, there is a different strategy for advertisers in *Art Business News* and *Art World News* than there is in *DECOR*. You see few ads in *DECOR* displaying a single image whereas it is often the case in the former two. The difference is in the price points of the images. If you have a $2,000 serigraph or mixed media giclée, perhaps you want to show it as large as possible, whereas the poster advertiser in *DECOR* may want to show more images with lower price points.

The tabloid page size used by both ABN and AWN create the opportunity to display an impressive single image, which is another reason why different strategies are used in different publications. You have to choose the one that works for your situation.

Regardless of what you may feel about aesthetics and the "art" business, don't allow them to get in the way of focusing the core reason for your ad, which should be to gain dealers. That means you should make it as easy as possible for them to respond to you. You should invite their response. This is not crass; it is good business and asking for response can be done in a tasteful manner that still satisfies a need to have a grand, eloquent presence if that is your goal.

Advertising rates should be called "advertising suggestions." That is, you should probably never have to pay the rates shown on a rate card. To start, there is the standard advertising agency rate discount that amounts to 15%. The rate was initially set up to help advertisers offset the costs involved in producing the ads for the publication.

Some publications balk at offering this discount to all but recognized agencies, which are those who are listed in some specific directory or other as an ad agency. This would be opposed to your in-house agency with no such listings — don't be dissuaded from getting this discount. If you are creating the ad materials and providing everything that an ad agency would provide in the specified format, whether digitally or in four-color separations with a proof, you should be entitled to the discount as an ad agency is.

When you see an advertiser's rate card, it is usually showing you what are known as Gross Rates. These rates are those before the 15% agency discount. When the rates are taken down by the 15%, they are called Net Rates. As an advertiser in a trade publication, you should always strive to work your deals off the Net Rate, not the Gross Rate.

Although publishers try hard to price advertising fairly, or at least competitively, they rarely, if ever have all their advertisers paying the same rate for their earned insertions. Earned insertions usually come in at 6, 12, 18, and 24-times within a 12-month period. Sometimes you'll see a 3-time rate. "Earned rate" means the advertiser has, for example, contracted to run 6-times over 12 consecutive months and thus has "earned" the 6-time rate.

The advertiser can choose the months in any sequence, six consecutive, on two, off two, every other month and so on, as long as they meet the requirement to run the agreed number of ads during the contract period. You can even combine a smaller ad with a larger one in the same issue and each would count the same towards your frequency discount. That's not a suggestion to go that route unless it truly makes sense for you; it is something worth knowing.

The point here is it is your responsibility to obtain the best rate possible. You should make an effort to get to know and develop a personal relationship with your rep. This can be advantageous to you in many ways. Reps are in contact with many of your direct competitors. They are hearing and seeing things that would be helpful for you to know. They are aware of opportunities long before you will be, and for all these reasons, they can help you in tangible and intangible ways to improve your business. It is in their best interest for you to succeed, but they are the same as everyone else, which means they are more likely to go the extra mile for those with whom they are comfortable.

If you have a problem with the publication, reps are the ones you will turn to first to straighten things out. I used to tell advertisers that printers and dry cleaners are similar in that if you give them enough chances, they will eventually screw something up. For instance, run your ad upside down, leave your ad out of the publication, put it in a place other than promised, and so forth. If you have a cordial relationship, you're going to get a better, quicker and probably more satisfying response than if you have an adversarial relationship. It just makes sense. And, your business ought to be worth a good meal now and then, too.

To some degree, ad reps serve two masters in that they both advocate for you to help you get the most from your relationship with the publication, but also are paid on commission and the more you spend, they more they make. Normal advertising commissions for independent reps run between 15-20% of the cost of whatever you are billed. Usually, they are paid between 5-10% for tradeshow space. The good ones are well compensated, which means you don't have to feel guilty about asking for their help to get a better rate.

While good independent magazine reps do make a decent income, keep in mind that they are not subsidized in any way by the publications they serve. Other than occasionally receiving comped rooms at shows and a few meals on the publisher's credit card, they pay all their expenses including health, retirement, travel, mail, office, and phone, and so forth to get the message to you and help the magazine stay viable. Reps that are actual employees of a publication are under different compensation, but to a large degree, they nearly always work on some sort of commission.

Having viable trade magazines should be important to you. They serve the purpose of educating your customers. By helping them become better and smarter operators of their businesses, they in turn help make them better customers for you. They help introduce new customers and prospects for you and they act as the voice of the industry in trying times when leadership is required. If trade publications ceased to exist, your means of getting to market would be sorely hampered.

Reps are loyal to the publication because that relationship represents their livelihood. Yet, they must also be loyal to their advertisers because participation and timely payment fund their livelihood. By understanding that dynamic, it will be helpful to you in negotiating with them.

Remember, rarely does a rep have the ability to make deals beyond a certain point, for example move a 12-time schedule to the 18-time rate without management approval. And, by asking them to help you negotiate and advocate getting a lower rate for your advertising, you are asking them to make less money. Therefore, having a good relationship with your rep and the magazine management is always going to help you get the best rate and deals on ancillary products.

I do not suggest opening with the following strategy. Nevertheless, if I felt I was not getting the best rate, I would use my hardest-nosed method of getting to the bottom line. That is where I would ask my rep to tell me straight forward what is the best rate that anyone running a similar schedule, for instance 6-times full-page four-color, is getting. If there is any hesitancy at all, ask them to put it into writing for you. Many people will shade the truth in conversation, but far fewer will do so in writing.

The negotiation should work this way. You, my rep, are asking me to sign a contract to complete a certain number of ads at an agreed upon rate within 12 months of the first insertion. I agree to this request. In return, I am asking you to warrant in writing to me that I am not subsidizing anyone else's lower advertising rate for similar schedules by paying more than they are.

While negotiations can become contentious, if you have established a rapport with your rep that shows you are honorable and you will complete your contract and pay on time, they should be agreeable to making sure you are paying the lowest rate possible for your advertising.

Fair is fair, hence you must be aware there could be other considerations included in a company's ad rate structure. They could be things as length of time a company has advertised and the amount of tradeshow space they take. It also might include their participation in other promotions, or the frequent availability of their ski chalet in Aspen, (okay that last one was there to test if you are paying attention).

If there are discrepancies that could cause another company with a comparable ad schedule to get a lower rate than yours, you should know what aspects outside their earned rate helped them lower theirs in order for you to make multi-year plans accordingly. Knowing the rules implied and otherwise, will help make you a smarter marketer.

Multi-year advertisers are the norm in trade publishing. These companies don't continue out of undying loyalty to the publication or the personnel. They do it because it helps them achieve their marketing plan. You should be thinking as quickly as possible to move from stressing out over making the payments on your ad schedule for one year to how your advertising will affect your business over many years.

I mentioned earlier about thinking about branding and direct response in flights of ads. (Flights of ads are a sequence with one building on the message of the previous.) Implementing a complex strategy can take years. You have a finite number of issues in any year to advertise in, even if you are in every publication. But thinking strategically for the long term and the short term will help you employ tactics that you might not otherwise have even considered. Don't be afraid to think in big terms or for longer than the next show season in conceptualizing and strategizing your ad campaigns.

The number of multi-year advertisers in art publishing trade magazines is impressive. *DECOR* for instance, has had advertising from some companies for more than 100 years. Its back cover was sold on a 12-time rate to the same advertiser for more than 60 years. Now that's long range! The other publications can't make that claim because they haven't been around that long, but they all have advertising from the same companies since their inceptions.

You don't need to be thinking in terms of decades for planning, but having the thought in your mind that you are in the game for 5 to 10 years is appropriate. It is how you will build a huge wealth of brand name integrity and many other attributes that suit your style to go along with your continuous efforts to build your dealer network through advertising.

You will find that many times publications will be looking to offer you deals on ancillary products in lieu of cutting your rate. Sometimes these will work well for you. The majority of them have a list, as mentioned above, of value-added items to offer advertisers. The list can range from preferential treatment in certain editorial or promotional features to links or banner ad on their Website, access to their mailing lists, reduced rate or free ads in their tradeshow directories and more.

You should take advantage of all the value added items available to you. As with publicity, you'll see that many of the smartest and best marketing companies are using everything at their disposal to help them get their messages out. The interesting thing about advertising in trade publications is that it levels the field. A $20 million publishing company's full-page four-color ad is no bigger than the same size ad from a new start up operation.

When you use this tremendous dynamic to leverage your position against much larger companies, and at the same time participate in all of the value added and publicity available to you, you are creating a mass of impressions and upping your visibility to the highest possible point.

Knowing all this is available, why would you not do as much of it as possible? Granted, it takes time and money to follow up and do these things, but if you include them in your plan from the outset, you don't have to try to shoe horn them into your operation later. From the beginning, hire someone whose responsibilities include taking advantage of all the value added and publicity offered to you. Use the savings on the premium ads to help fund this if necessary, it's that important.

By emphasizing value added, promotion and so forth, I'm not diminishing the importance of trade advertising. It is and will always be your best method of gaining new dealers. Just don't get so focused on advertising or a combination of advertising and tradeshows that you let other opportunities pass you by.

You don't want the sickening realization that other companies who enter into the industry around the same time as, or worse later than, yours seem to grow faster than yours and they are doing these things in concert with their aggressive advertising to help take market share from you. Don't let it happen, you don't have to if you keep the three "D's," Desire, Discipline and Details in balance in your business.

Tradeshows

Use tradeshows in conjunction with magazine advertising to form a powerful nucleus around which your marketing will revolve. If you will learn to make effective use of these two marketing options, and use publicity wisely, you will be well on the way to self-publishing success.

Tradeshows are wonderful because they are tactile. You can see the print or poster, usually well framed and lit, in person and that makes them important. Potential buyers get to meet you and you them. A problem with tradeshows is they are infrequent. There are scant few good national opportunities for tradeshows each year — not to mention they are expensive to undertake.

The nature of the business is ever changing. Tradeshows and their players continue to come and go. For instance, when the PPFA (Professional Picture Framers Association), *www.ppfa.com*, merged with PMAI (Photo Marketing Association International), *www.pmai.org,* it no longer produced its own stand-alone annual tradeshow or any regional shows. The PMAI still has a national show and PPFA is part of it, but it is no longer a strong show for art print publishers to exhibit.

Hobby Hill Publishing produces the West Coast Art & Frame Show, *www.wcafshow.com*, annually in early January in Las Vegas. They also are the publishers of *Picture Framing Magazine* for the professional picture framing audience. This has become the de facto show in the Western states since both Decor Expo and ArtExpo have both abandoned the area. PFM produced a show in 2004 in Philadelphia, but apparently, it has decided not to continue with one in 2005, another indication of the difficulty of producing profitable shows in today's market conditions.

A recent addition to the New York trade show scene is the Fine Art Forum. It came about when a number of fine art exhibitors who wished for a trade-only fine art pavilion at ArtExpo decided to start their own competing venue. FAF, *www.fine-art-forum.com*, runs concurrent with ArtExpo and Decor Expo in New York. It is a small show with some of the industry's leading print publishers from the middle tier of the art business on display there. This is not a showplace for new exhibitors, but being asked to participate is certainly something to aspire to for new publishing ventures.

Decor Expo, *www.decor-expo.com*, started out as the Art Buyers Caravan shows and used to produce six or seven regional shows in addition to its large New York shows, Galeria and Frame-o-rama. It now is under new ownership and in response to changing market conditions; the decision was made to eliminate regional tradeshows from the schedule. This show runs concurrent with ArtExpo at the Javits Center in Manhattan in late February or early March each year.

Pfingsten Publishing LLC, which gained *DECOR* and the ABC shows when it purchased the former Commerce Publishing Company in 1999, has been busy acquiring other competitors. In March of 2004, it stunned the industry with its announcement of the purchase of ArtExpo, *Art Business News* and *Framing Business Monthly* from its mammoth rival publisher and tradeshow producer, Advanstar Communications. This was not long after purchasing the European publication, *Art Expressions*, and re-branding it as the quarterly publication *Decor International*. And, it has added the aforementioned *Volume* to its stable of magazines.

Those acquisitions put Pfingsten into a formidable position of power by eliminating the chance to jump from ArtExpo, *www.artexpos.com*, to Decor Expo and vice-versa if you weren't pleased with the way the shows were produced. It is the nature of tradeshows to produce complaints from exhibitors, some justified and some not. Now, your choices of the companies that you can exhibit with in tradeshows are more limited than ever. That means keeping your cool and using the "To get along, go along," philosophy will help you more than losing it and stature in shows at the same time.

As others are, I am always concerned to see power concentrated, whether it's a few companies owning all the national media outlets or one company owning most of the magazines and all the largest and most important tradeshows in an industry, but that is the way of the world today. You can't change it; you can only learn to live with it.

Time will tell how the merger of these operations affects results for exhibitors and advertisers. If properly and fairly administered, it could lead to a boon, or if not a boondoggle. Another scenario to consider is that the ownership will change in the next few years as the nature of investment owning media companies is to buy, streamline, consolidate and sell. Pfingsten Partners LLC formed Pfingsten Publishing LLC and bought these shows in 1999 with the stated intention of selling within seven years.

There is a new tradeshow being launched in October 2005. The first annual edition of the Art & Framing SHOWCASE, *www.marketplaceexpos.com*, will run at New York City's Pier 94. It is expected to present nearly 200 art resources and framing suppliers and will attract 5,000 trade attendees, including art/framing retailers and volume art/framing purchasers. It will coincide with the existing "House to Home" Market, which runs in mid-October at NY City's Jacob K. Javits Convention Center. "House to Home" features 900 exhibitors from around the world in 120,000 net square feet of exhibit space and attracts 8,000 buyers. Product categories include home textile fashions, gourmet housewares, tabletop, decorative furnishings and accessories.

When it comes to any of these tradeshows, there will be lots for you to learn. The more you know before you go to your first one, the better the experience will be for you. Attending one or two in advance would be ideal. As a tradeshow representative, I used to poetically tell my first time exhibitors that their experiences would be as valuable for what they learned as what they earned. Naturally, most came with high hopes and anticipating great things, but still were unaware of many of the nuances that only experience can provide. They had to go through the experience themselves to understand what I really meant.

At shows you will find those around you with more experience can often become great resources and fountains of knowledge for you to tap. There are friendships to be made, too. Yes, you are all there to sell to the same people, but in this industry, you will encounter many decent, helpful people.

I recall meeting first-time exhibitors at the Pomona ABC show sometime in the late 1990s. They had stayed over for a week to come to the show. The previous weekend they had been at the Hobby Show and they couldn't believe the difference in how the exhibitors treated each other. At the Hobby Show, newbies were left to themselves and people were competitive and unfriendly. These exhibitors found the opposite at that Decor Art Buyers Caravan show only a week later and couldn't help but remark about it to me

As you get involved in booking your own show space, you'll discover there is an art to finding the right location for your company in the show. Many companies have strong opinions about where those best places are. As a new company without standing, you will have to get used to working your way up the priority list. But, it will greatly help your cause if you and your marketing people plan in advance to work hard at establishing a great relationship with the show management staff and, of course, your rep.

Booths are assigned according to point rankings with the shows in terms of years at the show, size of your booth, whether you advertise and how frequently. This means the companies with the most points get first dibs on the choicest spots. That's the law of the jungle at work and presumably fair. Nevertheless, you'll always find examples of companies in good spaces that don't seem to fit the qualifying criteria. If you investigate, you'll often find they have established a positive working relationship with those people who are in a position to do them some good (their rep and the show staff) when an opportunity arises.

Shows are dynamic things and booth assignments are changed for lots of reasons. Sometimes spaces open up that you can get that wouldn't normally be available to a newcomer. So keep in mind to work on getting to know show staff and your booth space rep, they can really help you if you give them a chance and a reason — work on gaining their friendship and respect.

A word to the wise...making unreasonable demands on the show staff is in the long run self defeating. If you want people looking out for your best interests, being an overbearing customer is not the way to do it. You can be forceful and polite at the same time. Under pressure it's harder, granted...but try anyway, it's for your own good to do so.

Having personally attended more shows as a representative of the show management than most people go to in a lifetime and having experienced bad behavior by people under pressure for all kinds of reasons, I can testify it rarely pays off. It's true you catch more flies with honey than vinegar in life, and it's true in tradeshows for booth placement and having special favors granted.

Of course, budget concerns dictate how much you can do and that is where the concept of setting realistic goals comes into play. You need to determine how much you can put into the business in terms of money and personnel time. This will give you the best chance to meet rationally set expectations.

When you establish unrealistic expectations because your heart overrules your head, you are setting yourself up for both headache and heartache. Don't do it! The best way to crush your business and your enthusiasm is to believe bigger things than possible are going to happen to you because you are special, different, lucky, star-crossed or blessed. Once the art is created, it's all about the marketing and the budget and how you allocate them.

Below is some advice proffered from Kim Klatt and embellished by me. Kim has been a leading tradeshow sales rep for many years. His list of clients and contacts and industry experience is extensive. Detailed below is a 3-Step advice plan for tradeshow attendees. Many of these items are invaluable and worth noting, especially for those of you who are ready to embark on your first tradeshow:

1. Pre-Show Activities

2. At Show Activities

3. Post-Show Activities

Pre-Show

- Determine Your Objectives For The Show
    - On site sales
    - Develop sales leads
    - Introduce new releases
    - Grow existing or open new accounts
    - Networking
    - Market research
    - Image building
    - Make plans for "Show Specials"
- Make A Plan Of Action To Help Your Objectives
    - Plan and plot the layout of your booth before you leave
    - Be prepared for unexpected occurrences, for things to take longer and be more expensive than you imagine
- Have a checklist for everything:
    - Business cards
    - Sales literature
    - Bags or Kraft paper for onsite sales
    - Office supplies (ink pens, stapler, staples, rubber bands, and so forth)
    - File folders

    - Clipboards for order forms
    - Order forms
    - Tape & tape guns
    - Copies of your contract and all paperwork
    - Address book
    - Candy & candy bowl
    - Punch or fish bowl for drawing
    - Medications
    - Extra glasses or contacts
    - If you think you'll need it, take it as to not waste time and energy tracking stuff down at the show
    - This list could go on for pages, hopefully these items will start you thinking of what is important to bring
- What do you want to communicate at the show?
    - Uniqueness of your art
    - New marketing plan for galleries
    - New artists for the company
    - New products or types of prints
    - What artists or images do you plan to highlight at the show?
- What do you want to accomplish at the show?
    - Sales

  - Sales leads
  - Possible new artist contacts
  - New vendors for framing supplies
  - New printer contracts
  - New sales reps
  - New ideas for marketing and selling
  - Impressions and suggestions for new art trends
  - Industry information and developments
- Pre-show promotions are crucial
  - Begin trade advertising 2-3 months prior
  - Begin publicity 2-3 months prior
  - Drop direct mail 2-4 weeks prior
  - Send slides, photos, samples
- Reasons Pre-show promotion is important
  - Your best customers and prospects are gathering at one time, they need to know you will be there
  - Most seasoned attendees plan their show in advance with a specific agenda — you want to be on it
  - Give your prospects your booth number and directions to the show if they are coming for the first time

  - Make time for social activities, dinner, breakfast, and so forth
  - Create awareness for "Show Specials"
  - Let certain buyers know that you are holding "special" pieces for them to see
- Staff Training
  - Product knowledge
  - Competitive knowledge
  - Sales techniques

At Show

- Arrange your booth in an open inviting fashion
- Don't block openness of your booth by putting a table between you and customers.
- Be inviting, you want buyers in your booth, not in the aisle looking in
- Creatively use as much space in your booth as possible without making it look cluttered
- Don't be shy about excusing yourself from salespeople, other exhibitors or anyone else who has your attention to talk with buyers who enter your booth
- Don't let boorish overbearing people command your attention when other buyers are come into your booth
- Be ready to easily capture contact information, name, address, phone and email

    - Give everybody something for giving you their information; mini-print, note cards, free shipping on their first order after the show; upgrade to overnight shipping one-time for the price of ground delivery
    - Use whatever ideas you have to be creative and have fun with your customers and prospects
- Arrive at your booth early
    - Walk the show to see your competition
    - Check lights and if everything else is okay
    - Make friends with exhibitors around you
    - Bring some snacks and drinking water with you
    - If it's cold/flu season, bring an antiseptic waterless hand lotion with you
- Make time to enjoy some of the city you are in
- Make a point of introducing yourself to the show and magazine editorial staff, it's probably the only time you'll meet them face-to-face
- Reasons selling at show are different from elsewhere:
    - Visitors are bombarded with information in a short time frame
    - Visitors are talking with your competitors at the show
    - Visitors are tired

    - Visitors give you precious few minutes to capture their attention as opposed to a 30-60 minute sales call

- Timing of the Actual Selling Encounter in Your Booth

    - Engagement of Prospect – 3 seconds

    - Prospect Profile Matching – 30-45 seconds

    - Qualification of Prospect's Needs – 3-6 minutes

    - Communicating your message and the next step – 60 seconds

- Have extra clipboards with order form in ready to give to buyers if you are busy — ask them to begin to fill in contact information and note the pieces they like, it will help keep them from leaving

- Think about having an unannounced be-back offer — if a buyer says they will be-back, give them an otherwise out-of-sight pre-printed "special" offer that describes your art, gives your booth number and is good during show hours

Post Show

- The three rules to post show activity are:

    - Follow up

    - Follow up

    - Follow up

- Be prepared to immediately enter all the names and addresses of those prospects and customers you met at the show. If you can bring a laptop and do it during down hours at the show, you are ahead

- Send everyone you met at the show a postcard, email or personal note

    - Thank them for coming to your booth

    - Offer an extension on your "Show Specials" for an extra 30 days

    - Call your hottest prospects instead of sending mail to them

- Don't wait, your competition is seasoned and they will be taking sales away from you if you don't follow up quickly

Since we are on the subject of sales, here's a quick take on how every prospective buyer can be classified into one of just three categories:

1. The first will buy from you with little incentive; they are already predisposed to your images. They are what are known as the low hanging fruit. You won't be successful if you only sell to these folks

2. The second is interested, but needs convincing and handholding. They could be trying to decide between your images and your competitors, or have other reasons for dawdling on the decision. If you master the art of selling to this crowd, I guarantee you will be successful. They hold the keys to your prosperity.

3. The final group is comprised of those who aren't ever going to buy from you for whatever reason. That's life; you can't win them all and shouldn't knock yourself out trying to convince these prospects to buy from you. The trick is to know when to give up on them.

As you learn to master how to effectively sell to the middle range buyers, you'll also begin to have a feel for when to give up on that range of buyers who really aren't your prospects. We discussed earlier how Best Buy had changed its approach to concentrate on the customers who represent the best margins and profit and realized a dramatic increase in profitability in a short time as a result. By learning how to close, convert, recycle or discard these three types of buyers, your business can emulate the strategy of a billion dollar retailer.

At some shows, ArtExpo is perfect example; you will be in direct contact with trade buyers and consumers. Depending on your business model, you will have to decide whether to sell to both, or to limit your sales activities to trade buyers.

As a new company, you need collectors and sales and to generate income when the opportunity strikes. It can be very tempting to sell to consumers under those circumstances. You will see established publishers and artists that sell to both at ArtExpo, too. At first, it's easy, you take sales where you can find them, consumers or trade buyers.

As your list of dealers grows, you will have to come to your own conclusion regarding selling direct to consumers. Because this situation affects many areas of your operation, you will find more discussion on this subject as you read through the book. Whatever you decide, your intentions on selling should be clear in your mind before you setup for ArtExpo or other shows with a mix of trade buyers and consumers.

Publicity & Promotion

Publicity and promotion are similar, but unique activities. Publicity can be construed as primarily the act of gaining exposure in magazines, newspapers, Websites and other media. Promotion would be personal appearances, giveaways, contests, et cetera.

When you combine the tremendous free publicity opportunities that are available from the trade magazines, blend in direct mail with your advertising and tradeshow exhibits and commit to some promotional activities, you are hitting on all cylinders. The way to round out the efforts is to have a Website and do regular email marketing to those names you have carefully opted into your list. We'll discuss e-marketing in a later chapter.

There are two components to publicity. The first is traditional publicity and that is the effort to have you, your artwork or something about your company mentioned in an editorial feature in some form of media as magazines, newspapers, Websites, television, radio and so forth.

While the first is critically important for you in creating awareness and synergizing with your other marketing efforts, the second is even more important. If you are an introvert, this suggestion might not be your fondest desire because it is for you to become "the brand." That is, your name becomes branded to a look and style that is unmistakably yours. To a degree, it is building on the cult of personality that is pervasive in U.S. culture. The quote below from decades long successful limited edition artist sums it up:

> I used to think if the art was good it would sell itself. Then I worked and starved for 15 years, and I realized that today's art business is about selling your name.
>
> – Steve Hanks

Selling your name is not selling out. And, by the way, what is selling out anyway? Does it mean you do ignoble acts in order to earn crass dollars to feed your family, enjoy your life and possibly do good works with your renown and excess cash? I think not — to those who would make such statements, I advise them to listen to this refrain from Bob Dylan's "Positively 4th Street." It was written in response to those former friends who turned on him when he plugged in at the Newport Folk Festival:

I wish that for just one time
You could stand inside my shoes
And just for that one moment
I could be you

Yes, I wish that for just one time
You could stand inside my shoes
You'd know what a drag it is
To see you

I'm never sure how much is plain jealousy or misplaced good intentions when I hear remarks about someone having "sold out." I would love to see the contract they signed when they did. And, even more, I would love to know with whom they wrote it. Perhaps some horned Faustian devil?

Did Elvis sell out when he became a lounge act in Vegas? Or was it much sooner when he traded being a rock n' roll pioneer for being a movie star matinee idol? Was it his job to defend some high ground for rock n' roll's elite? Who makes up these rules?

More on Dylan, he rejected the notion that he was the "Chosen One," the "Voice of His Generation" and many other unwanted labels people and even old friends and lovers including Joan Baez desperately tried to foist upon him. In his 2005 television interview, the first in 19 years, he told Ed Bradley of "60 Minutes" that he did not aspire to be anybody's Messiah, perhaps to be Elvis, but that was the extent of his longing for fame.

When Ed Bradley mentioned that Dylan's song, "Like a Rolling Stone" was named the best song of a list of 500 compiled by *Rolling Stone* magazine and asked if Dylan was honored, he nonchalantly replied, "Yeah" and agreed it was okay to get the notoriety. Bradley pressed him to admit it was an honor and Dylan agreed with a wry smile and said, "This week." Clearly, he knows taste is fickle and history is long and isn't concerned with what today's pundits make of his body of work — you shouldn't be worried about critics either.

In the end, you can only be true to yourself and if you are genuine in that, it will be abundantly clear to those who want to own a piece of your art. If it happens that thousands of people respond to your creative vision and willingly pay money to make it part of their lives, then you can be proud of your achievements despite what any critic might have to say about it. Perhaps Frank Sinatra had the best take on critics with this famous quote, "Living well is the best revenge." Henry Ford II, the grandson of automobile magnate Henry Ford, put it this way, "Don't explain, don't complain."

Regarding publicity, what I am suggesting here is if you incorporate branding your name into your marketing and particularly into your promotional efforts, you will have vastly increased success in selling your art. A story that sheds light on how thinking big and promotionally was in the mind of the most famous painter of the 20th Century, Pablo Picasso was related to me by Vince Fazio, Director of Education at the Sedona Arts Center. *www.sedonaartscenter.com* (By the way, if you want to visit one of the most beautiful naturally spiritual places you'll ever encounter and take some awesome art instruction too, contact Vince.)

A young painter had become acquainted with Picasso and made repeated requests for Picasso to view and comment on his work. Eventually, Picasso granted him the wish and went to look at his work. After taking considerable time in studying the young painter's work without saying a word, he finally spoke. His simple and succinct advice was this, "Sign your name larger."

You'll note that you never have to look to find a signature on a piece by Picasso. Why? He understood the power of his name. He might not have ever had a conversation with anyone about branding his name, nevertheless he knew, perhaps through instinctive genius or maybe through crafty wiliness, that to take pride in promoting his name would always work in his favor.

There is an interesting book on publicity and promotion called, *Get Slightly Famous: Become a Celebrity in Your Field and Attract More Business with Less Effort* by Steven Van Yoder. Although it is not written for artists, it contains many great examples of how people with certain knowledge and skills can use them to promote themselves.

It was mentioned earlier, and is worth repeating, that you should be a student of your competition; you also should become a student of the various trade publications available. Become a subscriber of all the industry trade publications immediately! Do it now! A list of their Websites can be found in the Resources section of this book.

If you don't have Internet access, get it immediately. Make subscribing to art trade magazines the first thing you do with it. The Internet access is not a luxury for anyone serious about competing in the art print business; it's a necessity.

You need to learn and understand what trade magazines offer their readers and advertisers and how they differ from each other. Find out everything about the free publicity opportunities magazines provide artists and publishers. Keep in mind, you don't have to be doing business with these magazines to be mentioned in their publicity columns, but it doesn't hurt either.

They also produce publicity in the form of free promotional editorial for advertisers. Make sure you are aware of any offerings and what their deadlines are. You can't count on your rep to always look out for your best interest with regard to you submitting your PR in a timely manner. While they will try, most reps are carrying a heavy responsibility to many advertisers, often selling multiple titles and tradeshows. Ultimately, you need to learn to take advantage of all they offer – make it your responsibility.

You can greatly amplify your marketing by mixing in an effective publicity campaign. For example, if you start with a schedule of six ads to run over a 12-month period, and you diligently apply yourself to use every publicity opportunity everything publications offer, you can easily double or even triple the number of exposures you will have to the market in that same time period.

Exposure is a key ingredient to success in this game. As you should already know by now, you are in the business of building a dealer network one at a time and your combined advertising, tradeshow and publicity are the primary ways you will use to begin to make inroads with the established buying habits of dealers.

According to an article in the *Los Angeles Times Magazine*, Howard Fox, curator of the Los Angeles County Art Museum, annually attends hundreds of exhibitions and visits 50 to 60 galleries regularly. He also gets letters, emails, slides and invitations to shows and to view Websites. He finds the incredible velocity and volume of the offerings makes it very difficult for the unknown artist to stand out with him. He believes what it takes, besides talent, is the slow water drip of frequent exposure that makes an impression.

While we are admittedly comparing getting the attention of one curator to thousands of art dealers, the process still is the same. The steady, methodical, patient application of every means possible to put one's message in front of a buying influence...drip, drip, drip away ...will lead you to success. Water will carve stone if it drips enough in the same place.

In trade publishing, the names of the columns change as do those of the editors responsible for writing them often enough that it can drive you crazy keeping up. By comparison, imagine how hard you will work to build an ad or get ready for a trade show to working on PR. You need to stay focused on being abreast of changes and publicity opportunities.

This is not one more thing to put on your plate, it is an important component of your whole marketing strategy and unlike everything else you do in that strategy, it's free, save the time and effort to write and send it. Working hard at publicity is a prime example of how an organized, driven marketing person can use the tools available to gain a competitive advantage over those who don't get it or get it and don't do it.

It is a smart idea to cultivate relationships with the editors of the trade publications. You'd be surprised at how few of your competitors make the effort. To this I say...duh! You should be thanking them. Their lack of foresight or laziness helps you garner more attention from an activity that requires a small effort for a potentially large return. Having editorial personnel become aware of your company will invariably lead to more publicity for you; why wouldn't you do it?

Considerately approach your editors. Find out what shows they are attending. Ask for a meeting if you are going to be there too. Let them know you respect their time and inquire of them how they prefer to receive materials, what their timeline and deadlines are, and ask for suggestions to help you maximize your publicity opportunities. Do this and you will have far more success than the average company that drops something in the mail or sends in unsolicited email. Ask them if there are story ideas that you might be considered for, do your digging and keep it professional and courteous to get the best results.

If you are seeing a repeat pattern of suggestion of befriending reps, show staff, management and editorial people you are right. This is not gosh golly gee isn't everybody great advice. It is from observations over the years about what worked best for those companies and individuals who wrung every possible positive offering from shows, editors and magazines. Because they worked as hard at building good relationships with people who could help them, they got far more in return.

As you begin to roll out your marketing and as your images gain awareness from consumers, you will find some consumers will contact you directly seeking to purchase your published works. While some artists do play both sides of the street and compete with their dealer base, it is a bad idea and I strongly urge you to resist the temptation. Often those contacting you are looking to get a discounted price from what they would receive from your dealers. This is a sure-fire way to lose dealers, or at the least, gain their contempt.

This is different than selling full retail during consumer days at ArtExpo, in my opinion. At that show, you have paid a pretty penny to participate and making some consumer sales can help pay the cost of being there. When consumers start calling from Internet searches or because they saw an ad or publicity item in a trade magazine, I think you should turn it over to a dealer and let them in on the transaction. There is no harm in following up with the dealer to find out what transpired. If it's a sale and they have to order from you to fulfill, that should be a great opportunity to ask them to stock a little more. Quid pro quo is the way to go.

Keep in mind you are in the wholesale business with regard to your prints. Your originals are a different story unless you have galleries representing you for them, too. If that is the case, you will find those galleries unappreciative of any efforts by you to undermine their sales.

I know there are as many bad gallery stories as there are bad artist stories. It depends on what side of the fence you are on, I suppose. I believe if you keep your commitments to the right gallery, you will be paid back in honest effort from the gallery.

Sometimes, despite everyone's good intentions, things don't work out. Don't be bitter, just move on and keep faith that good galleries who will do right by you are out there. If they are not working for you, start first by asking what you have done for them lately. Not to say you should be abused and misused, but that you should make the effort to try to make things better before moving on. If you haven't seen sales or have seen them declining in a gallery, call the owner and talk about it. Ask questions. Ask what you can do to help them. The tone of voice and manner of response will speak volumes if your antennae are raised.

If your own Website is used to direct consumers and collectors to your dealer base or galleries, you are doing the honorable, classy right thing for your galleries and yourself. If it's your firm impression a gallery isn't working for you, you need to get out of that gallery, but don't shoot yourself in the foot with the ones that are working well for you by creating sales conflicts with them by selling direct to collectors.

I'll offer some examples later of artists who sell direct to consumers and through dealers. Let's just say for now, that none of them used that business model when they were building their reputation and their dealer base.

A trend developing is that publishers are advertising and promoting their works to consumers to drive traffic to their gallery dealers. While this has happened regularly in other industries, it has not been a key component in the marketing of art to consumers. In the past, galleries were distrustful of sharing their customer names with publishers.

The April 2005 issue of *Art World News* carried a front-page story titled, "How Galleries Profit From

Publisher's Consumer Ads." Now that is a breath of fresh air. It signals that both publishers and galleries realize they need to be more cooperative in the marketing environment that includes the pervasive Internet. This is not to say there has been no cooperation in the past, but to note the level and intensity of it is growing.

A wonderful example of an artist promoting her work to consumers on behalf of her galleries is P. Buckley Moss. This concept is not new to her, but then many things she has done for years could be picked up by her publishing colleagues and put to good use. I'll talk more about her in the next chapter.

If you have a major announcement you feel is worthy of recognition beyond the trade press, you can submit a press release to publications and other news disseminating organizations through a broad-brush news release. For those of you who have a polished press release and need some help in disseminating it, you can use the free service of Non Starving Artists, *www.nonstarvingartists.com*. PR Web, *www.prweb.com,* is another site that will distribute your release free.

Of course, there are paid services available as well. E-releases, *www.ereleases.com,* is one example. For $399, it will submit your press release to its extensive database of newspapers and magazines, plus you can pick two categories from its list of industry specific outlets, which consists of trade publications. Its art category currently lists 71 organizations on the submission list. For $599, it will write your press release, up to 500 words, and do the submissions as well.

A site dedicated to artists, Art Deadline, *www.artdeadline.com,* has an extensive media database and offers members PR services. You can sign up for its free newsletter and get some of the current deadlines available emailed to you monthly, or you can a take subscription and enjoy access to all its resources.

Here is another PR service to consider using: *www.prnewswire.com*

Some other things to consider regarding publicity are:

- Effectively and properly applied, it will pay greater dollar for dollar dividends than any other marketing activity in terms of recognition, because of the power of third party influence

- Don't overlook it even if you are busy, it's too important

- From the onset, budget for someone to do it for you as part of his or her job function as you work on your business plan

- Hire a good public relations firm with art experience, if you can afford one

- As an artist, get to personally know editors and publishers even if your public relations (PR) person already has a relationship with them — that (PR) person might leave, but you will still be there

- Publicity is more important at higher price points, but should <u>never</u> be left off any marketing plan

Direct Mail

Direct mail is important and you should have a system of capturing contact information with the capabilities of using that data to follow up with people and promote to them at appropriate times. This is a critical function and the better the software and more thorough your follow through on building your customer and prospect files, the better and more effective a marketer you will become.

Many publications will rent you a list of their subscribers. Contact them directly to get more information. Some will do email and fax broadcasts for you. Because of Federal regulations and the Internet, using faxes for promotional purposes has faded.

There are other ways to acquire lists, including using mail list brokers. An independent marketing company with great email lists of art and framing retailers is: My Smart Marketing, *www.mysmartmarketing.com.* Its affordable services also include Website building and hosting. Or you can build your own easy-to-use, template-based Website and save on the setup. My advice is, if you are not Web savvy, let it build the site for you for the low setup fee of $99, including registering your URL.

An excellent time to use direct mail is before trade shows. If you layer timely advertising, direct mail and publicity around your tradeshow exhibiting, you give yourself maximum exposure to your primary buying audience at the right time. You want to get your advertising in at least one month before a major show and preferably two. Your direct mail should be timed to hit in 2-3 weeks before a show. If you have done your homework, you'll have all the publicity possibilities laid out on a matrix so you know when to get the needed materials sent out.

Postcards are a great way to affordably help kick off your direct mail program. They can be efficient and effective when used properly. My fellow *Artist Career Training* Art World

Expert, Martha Retallick, has built a whole business around using postcards for marketing. You can find out more about her on her site at: *www.passionatepostcarder.com.* As with me, she teaches Tele-classes through A.C.T. You don't have to be a member to take the classes, however discounts are given to members. Sign up on the A.C.T. Website to receive its free newsletter.

There are companies that specialize in printing postcards at low costs. One is Modern Postcard, *www.modernpostcard.com.* This company will help you get your message out to a your customers and prospects with affordable four-color printing. They also offer mailing and list services to help you complete your project with one-stop shopping.

To succeed with all these items requires planning. With regard to the publications and mailings, the earlier you complete your tasks the better for you. Get to know the editors and production personnel at the publications you use. If they know you and like you, they sometimes can accomplish miracles for you. Don't abuse the relationship in the process by being chronically tardy and needy.

Speaking of matrixes, what I mean by that is a master timeline calendar that shows you when you are exhibiting, or have other major activities that will benefit from a full on marketing effort as new product releases, catalogs, supplements and so forth. Once you have penciled in what it is you want to focus on, start looking at what marketing opportunities are there and back up your timelines to get started in enough time to finish your projects.

If it takes you a month to get the photography done and have an ad put together to your satisfaction, give yourself five weeks. If it takes you two months to decide what is going with you to a show and how your booth is going to look, give yourself 10 weeks. You should always plot your tradeshow booths on paper and actually construct them, if possible, to see how they look and make sure you have enough lighting and other requirements long before you pack up and ship your materials to a show. The last thing you want to do when you arrive at a show, especially the first time, is to have to run around looking for essentials that you should have packed and sent or brought with you. With the preparation, anticipation and travel, you are going to have enough stress. Don't compound it by not being well prepared before you leave home.

Email and Websites

Email and Websites are mentioned in the list of things that will help you get to market. Because of their growing importancc, I'll cover them separately in another chapter.

Reps

Reps are the final item from the above bulleted list of marketing avenues for self-published artists. For the purposes of this book, reps are independent contractors who engage manufacturers and suppliers to help them secure business and contracts from their prospect pool. In some industries, as in home furnishings or gifts, good reps, while sometimes difficult to recruit, can be worth their weight in gold. When reps are kings, you can't get a business off the ground without them, or at least, your path to success is severely hampered without them.

This is not the case in the art business. Yes, there are all kinds of artists' agents and representatives out there looking to help artists get notoriety and get them into galleries and show and so forth. And for many artists, reps of this sort are the difference between success and failure. However, do not confuse what those agents or reps do with the job you expect a print rep to do for you.

If you look closely, you'll find few reps are getting the job done for self-published artists. Sometimes reps come along that also have picture frame molding lines, but you will nearly always be an after thought. For them you will represent a way to pick up extra income after they satisfy their primary income source. As with any broad statement, there are exceptions and should you ever find yourself with a rep that qualifies as an exception, count yourself blessed.

It's not that bad people settle into the role of rep; it's that anyone who is good at the job soon realizes the margins are better in other places. With art, price points are not that high by comparison to furniture for instance. Plus, margins are tight. Reps initially interested in the field and good at what they do soon gravitate to better paying areas as gifts and decorative accessories and home furnishings, where they can make much more money for the same amount of effort. Bottom line, you or your marketing maven must be your own best rep.

You have read an extensive list of ways to get your work to market. If you are going to master using these resources, you have to have a means of maintaining a dynamic database. There is no excuse. Powerful computers are dirt-cheap, e-books are proliferating and any good businessperson wouldn't think of trying to compete without a good contact manager/database program to help them. It's not something you can put off until later. If you don't have these resources, you need to get them now!

You can find a good affordable database source at: *www.workingartist.com.* They offer a program for only

$99 that does most, if not all the things you need from a good software program for contact management. ACT!, *www.act.com,* is a best selling generic powerful contact manager program I have used with great success. Even Microsoft Outlook or MS Works will work (not recommended unless you already own either of them and are on a serious budget). For those of you who are geeky and have the time, Filemaker Pro7 *www.filemaker.com* is an excellent and easy-to-use intuitive tool to build your own contact manager and database programs to your exacting specifications.

If you are budget minded, you can't do better than a free service offered generously by the folks at Art Network through this link: *http://marketingartist.com.* You will need an Internet connection and a browser to use it. I have not reviewed it, so I can't tell you more about it. Personally, I prefer to maintain critical data including contact management information on a computer where I can back up the information and take it offsite. Of course, by using the free service your data is already presumably backed up offsite, but also out of your total control. This is a small trade off you will have to consider, but there's no arguing the price.

If you are in the market for other free software that offers programs to compete with Microsoft Office, check out: **OpenOffice.org 2.0**. It is an open, feature-rich multi-platform office productivity suite. The user interface and the functionality are similar to other products in the market as in Microsoft Office or Lotus SmartSuite, but in contrast to these commercial products, OpenOffice.org is like the best things in life, **free with no strings attached**. The Website is: *www.openoffice.org.* The software has been downloaded by more than 16 million people and has been translated into 30 different languages.

The days of getting by with a phone, a file drawer and a Rolodex are as gone as 8-track tape players. You need to invest in proper technology to help your business compete and grow. The time and money you invest in purchasing and learning technology will be paid back to you many times over.

Secondary Market

As the art print market evolved, the need arose for a means for dealers, galleries and collectors to be able to buy and sell prints that were no longer available from the publisher. This loose network of interested parties encompasses the secondary market.

There is no formal secondary marketplace per se. Instead, there are those broker/dealers who buy, stock and sell fine art prints, usually after the editions are sold out at the publisher, but not always. There are stratifications within the secondary market as there are in the primary market. You might see elite prints by masters as Chagall, Picasso, Rembrandt, Andy Warhol and others of that sort sold at high prices and exhibited at Art on Paper and other shows of that class.

You will find the line of demarcation in the secondary market is much more blurred than in the primary. This means many brokers are far more willing to trade in prints that some of their customers would thumb their noses at owning. But since many do not have showrooms or galleries, these brokers won't have to worry about offending anyone by having them see artwork that might be considered trivial by some.

You also will find broker/dealers selling art and collectible photography at lower price points from popular artists as Thomas McKnight, Robert Bateman, Bev Doolittle,

Erté and more. If you want to become acquainted with the secondary market, take some time to review the business card ads for their services in the back of *Art Business News* and *Art World News*.

At the peak of the offset limited edition mania, many publishers were forcing their dealers to buy prints they had no market for. More than anything, it was a way to force volume and control budgets. But what happened was many of those prints were dumped immediately at a loss on the secondary market by galleries with no buyers for the unwanted prints. The result was many artists saw the value of their prints actually decline in the process.

The hangover from that era is still felt today, and the secondary market is not nearly as robust and active as it once was. The good stuff, or at least the most popular stuff, still commands prices that are much higher than their originals. It really is a buyer's market and the effect of eBay on the secondary market has been acute and negative for some broker/dealers.

What artists want to see is a brisk market of escalating prices for your work on both the primary and secondary market. If you or your publisher hasn't flooded the market, you have a chance of seeing prices for your sold out editions rise in the secondary market. Keep this in mind as you progress in your business. Take some time to try to find out how many of the artists you admire have a following in the secondary market and what the prices for their pieces are. It will be an eye opening exercise for you.

The March 2005 issue of Art Business News featured an article on the Secondary Market written by their contributing editor, Joshua Kaufman. He is a lawyer who specializes in trademarks and copyrights and is referenced elsewhere in this book. He points out the there has been a rise of Secondary Market companies operating on the Internet that are not helping the industry.

The problem with many of these operators is that they pose as having relationships with artists and publishers and indicate they have stock available for all manner of prints when in fact they have neither. Kaufman points out the numerous violations of copyright infringement and other potentially illegal or actionable things that these companies are causing with their deceptive marketing tactics.

These companies can cause disruption in the marketplace, distrust amongst collectors and steal sales away from those who have the legitimate right to sell and market various images. In many cases, they are downloading images and copying verbatim descriptions and artist bios et cetera from legitimate suppliers. This article points out the not so pretty side of the art business and serves as a pointed reminder that not everyone can be trusted.

Use this information to help keep your wits about you when you deal with unknown companies. Ask for referrals and vigorously check them out. If your gut tells you something is wrong, you should pay attention to those signals.

> Experience is one thing you can't get for nothing.
>
> – Oscar Wilde

How to Not Profit from the Art Print Business

Before I close this chapter, let me spin a couple of cautionary tales of what can happen when unbridled enthusiasm and access to large sums of money gets coupled with crazy unrestrained, ill-conceived notions about marketing that lead to attempting shortcuts.

The timing of the first story is the late 1980s, before giclées came into existence. The limited edition offset format was "red hot." Bev Doolittle and other artists were making waves with edition sizes that were astronomical. Doolittle's previously mentioned "Sacred Ground" piece sold nearly 70,000 copies and in the process set a standing record for limited edition size, which at that high number is difficult to describe as limited.

"Sacred Ground" was what is called a time-limited edition, meaning dealers had a specified time to get pre-orders for the piece and whatever the total number was when it closed determined the edition size. At $300 retail, that meant nearly $21 million in sales for those dealers in a short time — and all in advance. Wow! Those were the days, my friend.

Publishers seeing these astonishing numbers being sold rushed to get in on the action and their hype and the mini-mania of collectors, who were buying one for the wall and one for under the bed for speculation, followed them into the fray. As with other manias, things were too good to last.

Many collectors who bought less popular pieces, and even more dealers who were forced to buy the dogs in a line in order to get the really hot images, were eventually left holding the bag with lots of once expensive paper they couldn't give away.

Interestingly enough, the cream manages to rise to the top even in a deluge of too much product. The Doolittle "Sacred Ground" piece is still sold on the secondary market and eBay commanding $600 - $900 framed, which is double, or triple its original unframed cost. This is against a backdrop of a glutted market where many limited editions produced during that time are worthless on the secondary market.

The first story involved a fellow who called me from out of the blue one day and inquired about advertising and tradeshow rates. As with many calls of this nature, he had many questions and wanted to learn as much as possible. I eagerly answered his questions. He was keen to learn and I was prepared to help him do the best with what he had.

I won't divulge names, but the caller was a professional, and apparently a successful real estate agent. He had come across an artist whose work blew him away. The artist was from Uganda and was considered at the time one of the best artists in the country, if not all of Africa. The real estate agent was taken with the work of this artist and offered to start a publishing company to sell his images as offset limited editions.

With the limited edition mania still in full bloom, my new customer had his version of sugarplums dancing in his head. He took my advice, and started advertising and was getting modest results. He exhibited at one of the *DECOR*-sponsored regional ABC tradeshows and had again modest results. Keep in mind, he was pioneering a new artist and a genre not yet crossed over to mainstream.

He decided too many of *DECOR's* readers were not ethnic and were not as enthusiastic as a demographic with more Blacks would be. That may have had some minor validity, as the breakthrough of African-American art was still a few years off (remember that being too far in front of trends can be as bad as lagging them). The reality is he was impatient for success and unwilling to take the time to grow his dealer base one at a time.

As you can imagine, his eagerness was bad for him and the artist. While attempting to totally bypass the traditional means of getting to market, he hatched the following plan to jumpstart the art sales, figuring they would go off the charts. He took out a second mortgage on a piece of rental property he owned and borrowed around $35,000. At that time, this was more than enough money to buy an annual contract of monthly full-page ads in *DECOR* and exhibit at all the important tradeshows during the year as well.

Unfortunately, against my strong advice, he contacted *EBONY* magazine and found out at that time it had about a one million circulation or subscriber base with three million readers making up what is called pass-along readership. He figured by going direct to the consumer even if he got only a small fraction of a percent to buy, he would be on easy street.

He calculated that 1% of 1% of the 3 million readers would get him sales of at least 300 prints. As I recall, the retail price was $300. He figured he would gross $90,000 per ad.

Sorry, but believe me, if it were that simple, you wouldn't be reading this book. I would be skiing in Gstaad with the glitterati and this book would never have been written.

I don't recollect exactly how much the ad cost him, but I know it was most of the $35,000. I begged him to reconsider. It looked to me like a disaster in the making, but he would not listen and bought the ad. You already know this didn't work out for him, and you're right. It was a train wreck. He sold a small number of prints, failed to cover his costs and I never heard from him again.

The artist surfaced a few years later with a new publisher. They were attempting to reinvigorate many of the same prints originally published by the real estate agent and which by then were passé. The new publisher also was under-funded and trying to make a comeback as well. Neither of them ever had a chance.

What happened? Because his artist was African and the advertising was appearing in a consumer magazine that mostly reached African-American readers, there was no predisposition on the part of those consumers to buy art advertised in *EBONY.* It would be no different from the readers of TIME to pick up the phone and order limited editions prints from an unknown artist – even a talented one. How many ads for fine art prints have you seen in large circulation consumer magazines? It just doesn't work that way as that real estate agent painfully learned.

The artist's career suffered and he never got the reward and recognition he might have if the real estate agent turned art publisher hadn't got greedy, been impatient and acted irresponsibly with the artist's career and his own money. If you have any zany notions like this one, please do yourself a favor and avoid such tomfoolery, I beg you...Bora Bora beckons.

As bad as that was for the real estate agent and the artist, I witnessed an even greater amount of money lost on unfocused marketing and poor decision making. Again, no names are provided.

In this example, a wealthy venture capitalist and a talented artist became friends during the years of their children's sports activities. At some point, the venture capitalist decided to invest in the artist's career. The idea was to catapult him to recognition at a national level by marketing etchings in a series of themed collections.

Initial strong plans for multiple themed series editions, tradeshows, personal appearances, an artist curriculum vitae book and advertising were developed around the expectation the artist would produce the etchings on schedule.

This artist already was successful. He had built his quiet career earning a six-figure income from a niche where he had little competition. He benefited from the wily tactics of his shrewd rep (you see, there are exceptions) who had been with him for years. What started out with great promise ballooned into a mess when the financier brought in a decision maker with no art industry experience. Perhaps that this artist was successful made him less hungry than another artist might have been.

This dream project started with everything in place: a talented proven artist; excellent deep pocket financing; a well-conceived marketing plan; and a dedicated team of experienced art sales and marketing pros. A problem arose when the financier's handpicked person who had no art experience and little ability to control the artist took over. He wanted to be sure his money was watched closely. It didn't work.

Soon the artist's rep had no word in the decision making process and the whole thing went topsy-turvy. The untamable and unfocused artist began pursuing a separate goal of creating miniature sculptures for the gift and tabletop market. With the once aggressive goal for the company to receive new art prints falling behind schedule, the company switched strategy from etchings to giclées. This was a serious downgrade in the marketing plan.

At this point, it was too late. The whole venture came unraveled with one giclée print produced, a couple of ads in one of the trade publications and one tradeshow experience. The bottom line...the venture capitalist lost hundreds of thousands, and many other people spun their wheels, either getting little or nothing, as they worked on the bet that the promise of the outcome would be reached. The artist went back to plodding away in relative obscurity having squandered a magnificent chance to elevate his status and art to new levels of awareness and sales.

The sad thing is the failure could have been avoided by making sure the artist was committed and willing to stick to the original plan. The point I am making here is the plan you develop must be based on solid business and marketing principles with an artist committed to the plan. Keep that in mind as you move forward. Know what you want and stay tightly focused on getting there.

Ads are the cave art of the twentieth century.

– Marshall McLuhan

Chapter Seven

Some Exemplary Successful Self-published Artists

Art is never finished, only abandoned.

– Leonardo da Vinci

When you study the most successful self-published artists, you find the majority of them use a combination of all the possible ways to market their images. You also will find some who are successful despite the fact they choose not to incorporate all of the available elements in their marketing schemes. This means that if you execute well in enough areas, you can forgo an element or two and still get acceptable results.

I always recommend using every means available because I believe it is best to make as many positive impressions on your target market as possible. While I can't deny the success of some artists and publishers who avoid tradeshows or advertising and so forth, to me it doesn't make sense to exclude proven methods of marketing, communication and promotion.

Regardless of whether or not successful self-published artists use all the marketing tools available to them, I am confident they all have in common what I call:

Self-publishing Essentials

- Art that resonates with the buying public
- A business plan to create and sell the art
- A dedicated marketing/sales maven to make the business run

- An artist willing to prodigiously create in the same genre or theme
- Financing
- A clear understanding of the primary business they are in

Coincidentally, one of the artists covered in this chapter is one who no longer exhibits at art tradeshows. She is P. Buckley Moss. I refer to her several times throughout this book, and she is my heroine. Honestly, I have many marketing heroes in the print art business, but her story is compelling and needs to be told.

When it comes to the all important aspects of acting on knowing what business you are in by building and nurturing a dealer network, no artist has done it better than P. Buckley Moss. Others will have made more money or have more dealers, but I believe none have earned the loyalty of dealers and readers alike as she has.

Her level of personal involvement with her dealer network exceeds that of any artist I know of. Although I have never met her, I would consider it an honor and privilege to do so. Even though she has been a monthly advertiser in *DECOR* for far longer than my 15 years there, the opportunity to meet her never materialized.

Although I was never responsible for a sales territory with her company in it, I worked closely with colleagues who knew her from her early years and who had the chance to view her development. From them, I learned much about her including that she initially did exhibit at tradeshows. As her dealer network grew, she put her time into them instead.

What did more time with her dealers mean? In her case, she is said to have done more than 100 appearances for one-woman shows at her dealer galleries a year — and she did this for more than 15 years running. You can imagine with that much travel there would be no time left for tradeshows. For her it has paid off in a level of loyalty that is as strong as any artist has ever enjoyed, and that has led to fabulous sales results, too.

An example of how well her efforts produced for her galleries is in this story. When I worked for *DECOR*, I spent many years at the offices in St. Louis. One morning I opened my copy of the *St. Louis Post Dispatch* to read with great interest an article about P. Buckley Moss. She had come for a show at one of her dealers in Collinsville, Illinois, a sleepy 16,000-population bedroom community 12 miles across the Mississippi River from downtown St. Louis.

The article reported that collectors had lined up and Pat, her first name, signed prints, posters and collectibles for eight hours. The result was about $100,000 in sales in one day for that gallery. Do you think the owner of that shop would walk through walls for this artist? I know I would if she produced those results in my gallery.

The article reported that since Pat has dyslexia, she donated a part of every sale that day to local charities, and every show she does contributes help to children with reading disabilities. It's another way she connects with her audience and uses her celebrity to make a difference.

It is not just being there personally for her galleries. She and her marketing maven husband have had the most regular, consistent and ambitious advertising schedule, in both consumer and trade publications, of any print artist for more than two decades. I realize a demanding travel or advertising schedule is not possible for many artists, but you can still learn much from understanding how Pat Buckley Moss has created a loyal following among her dealers and collectors.

Since Pat and her husband have a definable niche, they use that to their advantage and advertise in the lifestyle consumer magazines that Pat's collectors are most likely to reading, such as *"Country Woman."* By doing that, she is helping her dealers by creating awareness for her work and driving sales to them. It is another way she helps herself by helping her dealers.

To my mind, there is no dichotomy in applauding Moss' advertising in "*Country Woman*" and reproving it as an idea for the real estate would-be art publisher. Once you have established yourself with a viable dealer base, the idea of branching out to help your dealers by advertising to drive traffic to them is applaudable. But, to attempt to circumvent building a dealer base and go directly to consumers with an expensive one-shot ad is career suicide.

If you can emulate the way P. Buckley has worked to build a dealer network and keep it enthused about you and your art, you are on the way to certifiable success. It's not easy, but little in life worth achieving is, wouldn't you agree?

Perhaps you can't spend that much time away from home to be able to support your galleries with personal visits. Then you have to start thinking creatively about what other ways you can support them to encourage them to sell your art when customers are in their galleries.

You could be thinking to yourself, "How can I do that? I have so many other obstacles and things going on in my life to be able to concentrate energy on my art business." If you tell me that, I will believe you. I also will tell you there are artists who are achieving enormous success in the art business today and they have overcome greater obstacles than what many artists have faced in their lives.

A prime example, and another hero for me, is Cao Yong. You can check out his vibrant international city, sea and landscapes at: *www.caoyongeditions.com.* Cao is a Chinese immigrant who came to the U.S. in 1994 and couldn't speak English.

Although he previously had a successful artist career in China, learning a new culture and a new language is still a huge hurdle to overcome for anyone. Can you imagine moving to China to make a go of it with your art career? The reverse is exactly what Cao faced. You can read his fascinating biography on his Website and learn even more about the many severe hurdles and hostilities he faced before he got to the U.S.

If you do research his site, you'll see he worked with a couple of the better publishers in the business (Aaron Ashley and Fortune Fine Art nee Colville Publishing) before emerging on his own as a self-published artist. He's a prolific painter, which is obvious by merely noting the output of his various series on his Website.

On his Website, you'll see he completed 14 images in 1998 – 1999 for the Golden Coast series. Of those 14 pieces as of March 2005, seven of the editions were sold out. If all seven were from a 350-edition size and they sold at retail for $1,000, which is probably a low price, that's nearly $2.5 million in retail sales for those seven pieces. And, he's sold a lot more art than those seven pieces over the years.

Kudos to Cao for his ability to tap into what his collectors want. It's something you want to aspire to in your career. You can learn more by studying his Website and that of other artists. It's always a good idea to try to find out as much as you can about how other successful artists are handling their work and Websites are often a great place to start. You can see Cao works in serigraphs and does giclées. And, in some cases with the giclées, he is selling the same image in two editions at separate prices — more concepts for you to ponder.

The lesson here is you should be a student of your competition. I guarantee if you start selling well, you'll have many people studying what it is you are doing to be successful and some will be trying to copy you.

A word about knockoffs; some knockoffs are blatant and obviously illegal. Others are borrowing your "hot" theme or look and incorporating it into their line. While the latter will infuriate you, get used to it. It is a part of the business. It's also a way you will know for sure you are in the big time when you discover you are being knocked off. There's a fine line between "creative borrowing" and outright copyright infringement. I'll tell you later on how to protect your intellectual property rights.

> Copy from one, it's plagiarism; copy from two, it's research.
>
> –Wilson Mizner

You'll see from Cao's timeline in his biography that he has become a regular at ArtExpo. I'm sure he finds the effort and expense of being there rewarding. The visibility dynamic in this business is extremely important. Again, I urge you to make the trip to New York if you want to see the art business up close and personal, and the works and marketing efforts of Cao Yong and other top-selling artists.

Another popular artist you might encounter when you attend ArtExpo is Yuroz, *www.yurozart.com.* He has a distinctive modern style that some say is a derivative of

Picasso. He is perhaps the poster boy for what happens when a serious fine artist is considered too commercial. This occurs when museums and institutions that would ostensibly ensure the reputation of an artist long after he has sprung off the mortal coil have chosen to ignore him.

You can read an incisive and well-written 4,100-word cover story from the February 2, 2002 issue of the Los Angeles Times Magazine titled, "*Never Mind the High Praise. How About a Little Ink? His Work Is Priced as High as $150,000. He's Been Commissioned to Paint by the U.N. But There's No Place in the World of Fine Art for Yuroz and Others Like Him*," by David Ferrell.

There is a small fee to download the whole article on the LA Times Website. The article offers a thoughtful view of this artist's work and his successes and anguish over being ignored by the museums he wishes would collect his work. I highly recommend reading it as part of your art print education.

The article title just about says it all. Here is an artist who has enjoyed tremendous success, one who was dirt poor and homeless. As with Cao Yong, Yuroz is an immigrant to the U.S., originally from Armenia. While he has enjoyed success many of his contemporaries would love to experience, for him it is bittersweet. As the article says, in his case and for his desire to be collected by museums, he probably overstayed at ArtExpo having been 14 years running at the show in 2002 when the article was written.

The gist of the article is that Yuroz, by taking the path of making an extraordinary income to create a great life for his family and make them as comfortable and safe as possible, could have let the chance at art immortality pass him by. But he also had no guarantee he would have garnered the museum attention he covets if he had been less commercial either.

It's a crapshoot when you give up the certainty of selling well at Artexpo for the uncertainty that the museum crowd will covet your work. And even artists that are considered "hot" by museums at one time may not stand the test of time. Julian Schnabel is an example that dynamic. Who knows what history will have to say about him, certainly no one living now can say.

I think that is what Dylan was indicating in the conversation with Ed Bradley, mentioned earlier in this book, when he was reluctant to be excited about his song topping a *Rolling Stone* magazine list of the 500 best songs.

Will you have to make such a choice? Since the odds for making it in the art business are stacked against you, most struggling artists should be so lucky. But you didn't choose to be an artist because it was the road more traveled; it's more like it chose you and you have followed it. However you found yourself on your path, it doesn't mean you have to ignore the chance to prosper from being on it.

Another way of making the difference between those artists who, like Yuroz, could be earning huge amounts of money and those who are being vetted into museums, often without large incomes, is characterized in an August 30, 1999 article in *Time* magazine titled, "The Art Of Selling Kitsch. Don't Look For These Creations At Your Local Museum. Instead Try Your Local Mall." For a small fee, you can download it from the *Time* magazine archives.

Here's part of the first paragraph from the article:

> "Seven years ago, struggling artist Thomas Kinkade sat in a secluded gallery well past closing time, determinedly propounding the virtues of his luminescent garden-and-cottage scenes to a young couple...He...told the couple...he was going to give it a few more months and if he couldn't...earn a living...he'd close up shop and move on."

According to the article, Kinkade went on to become the top earner of more than 30 palette-to-paycheck artists (as it described them), and who are multimillionaires and the opposite of starving artists. In the process, they have earned the scorn of art historians and fine art galleries who characterize these artists as populist artists.

I don't know about you, but I am not ashamed to admit I like a lot of the work that these artists (top selling print artists) create and I don't care if it is populist as opposed to museum quality or not. Still some of the stuff that is passed off as fine art astounds me. Despite all the rationales sycophant critics and art elitists come up with about the multi-layered deep meanings of it, some it escapes my interest and comprehension.

Given the choice between a pile of broken glass, a solid blue canvas and a Steve Hanks image of children languishing on a dock and I'll take the latter every time, as will most of the potential buyers for the art painted by most readers of this book. So, take heart and paint on!

Art is really about your choices and the choices of those whom would be your customers. Once you have zeroed in on whom they are, you can quit worrying about those outside your target range, even if that means ignoring museums, critics and Manhattan galleries.

The *Time* article ends with a poignant quote from Wyland, one of the top selling limited edition artists of our time.

> The art snobs frown on any marketing or business, but the old masters weren't successful until they were dead. I didn't want to wait that long.
>
> – Wyland

Where you come down on this debate of being too commercial, or getting in a museum, is a purely personal decision you have to make. I believe that as with the trio of choices I opened this book with regarding being full-time, part-time or hobbyist, there is no wrong choice here, only the one right for you.

Wyland might not get into any museums of note either, but that hasn't stopped him from gaining tremendous exposure for him and for his cause of creating awareness for and preserving our oceans and marine wildlife. What he has been done through his whaling walls program is remarkable, astute and amazing. His goal is to paint 100 murals depicting whales and other marine wildlife on buildings worldwide by 2011.

The whaling walls are a stroke of promotional genius. You can see some stunning pictures of him at work on these murals on his Website, *www.wyland.com*, where he claims that more than 1 billion people have viewed them. Think about the price if he had to pay a publication or television station to reach that many eyeballs. His whaling walls personify the best in publicity, in thinking big and acting on the thought.

The whaling wall promotion works on multiple levels. It creates awareness for Wyland, for his art and for his cause to work to raise the consciousness of mankind toward the ocean and the wildlife in it. With these magnificent murals, through which he has set and broken his own record for largest ever, I'll give odds he has done more to further his cause than nearly any *soi-distant* fine artist has ever done.

Wyland has become a multimillionaire along the way, (the *Time* article pegged his earnings as a $50 million business back in 1999). Meanwhile he enjoys what he is doing in bringing joy and love of the ocean to people. Who is to criticize his work and more importantly does he care if art snobs and critics ignore him and *Time* magazine calls him a populist artist? From my encounters with him, I'd have to say I'm sure he doesn't care.

Another artist whose work I admire greatly and who transformed from a self-published limited edition artist to selling primarily open edition prints was Richard Thompson.

He was a great American Impressionist artist of the 20th Century. He had the misfortune to be born at a time when Abstract art and Expressionism were all the rage. He passed away in his 70s in the mid-1990s after having created some 1,500 beautiful paintings in his lifetime. As with many of his generation, he started out his career as an illustrator in a time of Underwood typewriters, block type and zinc engravings.

The son of a painter, Richard Thompson, (find his work here: *www.impressionistprints.com* and *www.thompsonsfineart.com*) embodies a long line of illustrators who crossed over to find themselves pursuing their true passion in the field of fine art. The history of art in the 20th Century is laden with examples of one learning and working the craft of an illustrator only to use it as the foundation for another means of creative output that would reach people on a different plane. Some extraordinary artists who used the print medium to reach large audiences came from the role of illustrator, a nearly lost art today's Mac and Photoshop environment.

Richard Thompson's splendid originals regularly sold for $30,000 - $60,000 and higher, with some selling for well into six figures. In his lifetime, he was featured a record number of times on the cover of JAMA (Journal of American Medical Association), which is well known in art circles and by art aficionados as prestigious as getting on the back cover of Readers' Digest. Recognition from either of these publications is a huge career booster and something you might aspire to yourself.

Richard had the good fortune to have a son and daughter-in-law and eventually grandson, who loved his work and had the ability and desire to want to promote it. Together they built a terrific family business they operated out of the Maiden Lane gallery just off Union Square in San Francisco for many years. Entering it gave a person a glimpse of visiting a gallery filled with Monet paintings at a time when the master lived.

In addition to selling his original works from the gallery, they also formed a publishing company that initially sold offset limited edition prints of Richard's work. This was also during the aforementioned time when this kind of print reigned, before the introduction of the giclée format and the bursting of the offset limited edition bubble.

It has been said artists are more sensitive than others and by observing them carefully; you can get a foretaste of the future. I think Richard augured the decline of the limited edition bubble when he told his son as the company was doing well with the format that he no longer wanted to take the time to sign limited prints. He'd rather spend the time painting. How prescient was that for him? It helped his publishing company get ready for the next wave of marketing that would sustain them through the downturn of business in the limited market.

Richard also wanted more people to be able to see his work and would rather sell his prints as open editions allowing all who wanted them to be able to purchase them. His masterpiece, his best selling image and highest priced original are all one in the same, "Bellingrath Gardens." You'll find a beautifully framed print of it on a wall in my home. The print sold in the tens of thousands. Had it been a limited edition print, only a few hundred people would have ever had the great pleasure of owning and enjoying this print. I am one grateful collector that Richard made this decision.

His son had a great marketing mind and as a former ad sales rep, he knew the value of advertising and used it well over the years to promote the line of images. They also worked many of the ABC tradeshows on the Decor circuit and effectively used their direct mail lists. In many ways, their company was a model to emulate for those starting out. Today, his grandson and daughter-in-law continue to run the business after the unfortunate and early demise of his son, Richard Thompson, Jr.

There are countless stories of self-made self-published artists. They are far too numerous to mention in this book, much less attempt to provide anecdotes about them or their business models. But no matter how many of them you study, they usually encompass those things I set out saying were critical to the success of the mission in Chapter Four. If you need to refresh what they are, I encourage you to do so in order to keep in mind what is important to your business success.

Arnold Friberg, *www.fribergfineart.com,* who created a series of monumental works that toured every continent in the late 1950s to help promote Cecil B. DeMille's epic motion picture, "The Ten Commandments," is another good example of an artist whose career began as an illustrator. N.C. Wyeth, Norman Rockwell, Bob Timberlake, *www.bobtimberlake.com,* and Andy Warhol also fit in this category.

Perhaps Bob Timberlake is not a household name to some, but to the legions of collectors throughout the Southeast, he is a revered artist. His influence and style was famous long before Thomas Kinkade was licensing his look for La-Z-Boy Furniture, which by the way is no longer the case.

Timberlake's art propelled him into creating and licensing all manner of home furnishings in what could only be called manly Americana pieces, many taken directly from his cabin in North Carolina. They have been painstakingly recreated and sold with his name licensed on them by Lexington Home Brands.

Lexington, at their factory showroom in High Point, North Carolina reconstructed his cabin to display the whole line. When I visited there some years ago, I felt I had entered a time warp at Disney World. The complex housing the showroom is a modern steel and glass building. However, when you went up a flight of stairs, crossed the wooden bridge over the gurgling stream to his cabin, and heard the crickets chirping, you quickly lost that exterior modern perspective.

It was fascinating to see the extended reach and influence one creative person can have over so many areas that stretch well beyond art on paper. So, no MOMA shows are likely for Bob Timberlake, but no regrets are likely either.

Successful self-published artists come in all stripes. One, who crosses over many boundaries in the business, is from the area I would consider decorative arts, but nonetheless a print artist with a fine career. Her name is Mary Engelbreit. *www.maryengelbreit.com.* She is from St. Louis, and I confess had I not lived there for many years I might not have ever known about her. That is not to say she has a small following. Quite to the contrary, she has become a lifestyle maven and has her own magazine with a half a million subscribers. That's certainly no small potatoes!

The kinds of images she creates are cute, colorful, whimsical and homey, which is not what would catch my eye. Yet despite a personal lack of interest in her style, I still find it fascinating that the art business is so vast, with so many entry points. Here was a visionary leader in her field, a licensing phenomenon doing $100 million in annual sales, and she was off the radar of an otherwise seasoned and fairly sophisticated art marketing type, namely myself. Nonetheless, that was the case with her and me. Our spheres of influence might have never crossed were it not for her retail store in a suburban mall in St. Louis and occasional coverage about her in the *St. Louis Post Dispatch* daily newspaper.

Perhaps the reason why she was off my radar brings up an interesting thing about her. She bypassed the traditional means of getting her artwork to market, including all the concepts that I preach and that form the basis of this book. She doesn't advertise in any of the art and framing trade publications or exhibit at any of the shows. She started out in greeting cards and grew her art publishing and licensing business in the gift industry. Whoever reps *DECOR* in St. Louis is missing a great chance to help her open a new market. At $100 million in sales, it's not like she needs it.

So why would I go to the trouble to write about her when she flies in the face of my best advice and succeeds tremendously? Glad you asked. Two reasons come to mind:

1. To remind you that this book, while definitive when it comes to outlining the traditional means of building and marketing a print art career, can't cover every contingency

2. When someone is this successful, it's worth noting

There are many other licensing phenoms out there that I can't put into detail here, but who enjoy enormous success. A few examples include Flavia Weedn *www.flavia.com,* Jody Bergsma *www.bergsma.com* and Stephen Schutz *www.bluemountain.com.* Stephen Schutz and his poetry-writing wife, Susan Polis Schutz, have since sold their business, but no one can deny the tremendous success these artists have had with their careers.

It's certain that none of them are bound to find their works courted by topnotch museums, but I doubt that is really a concern for any of them. Success unconditionally is, after all, really in your own description. And to get back to Mary Engelbreit, her success is undeniable in terms of reaching people, using her talent and making great use of her creative abilities. That she is a plucky person is personified by a couple of quotes that can be read on her Website, the first in describing how she launched her own greeting card company when she was eight months pregnant:

> Proper timing is overrated. There's always a reason not to do things – it's too expensive, or it's not the best time, or this, or that – but I believe there are wonderful opportunities sailing by, and you have to be ready to grab them.
>
> – Mary Engelbreit

Her second quote neatly and succinctly wraps up the sentiment from a true American success story. If through reading this book, and much more importantly, by believing in yourself and applying your talent, you find you someday are living your dream, I will be proud if through something you learned in these pages had a small hand in guiding you there:

> I believed in myself and now I'm living my dream.
>
> – Mary Engelbreit

Chapter Eight

Protecting Your Intellectual Property Rights

> There is no short-cut to art, one has to work hard, be open and flexible in your mind, keep the child alive inside you, and through a whole lifetime be ready to learn new things and, of course, be mentally prepared for a hard punch on your nose — especially when you think you are doing well.
>
> – Bente Borsu, Actress

All you need to know about copyright protection cannot be covered in one chapter. If you aren't already aware of how to protect your intellectual property rights, then this will guide you to the places where you will find the information.

There are whole books on this subject and lawyers who make a living giving advice and creating legal documents to protect intellectual property rights for individuals and companies. Some of the biggest legal fights carried on by Microsoft, other software developers, multi-national corporations and the entire entertainment industry are centered on the efforts to keep rogue nations and individuals from pirating and abusing copyrights.

Of course, the fight is closer to home with the recording industry's RIAA and the motion picture industry's MPAA organizations aggressively taking on the freeloaders who illegally download music and movie files from each other. Some of the worst abusers are now being dragged into court and face stiff fines and even jail sentences for their actions.

These powerful associations have successfully sued Internet Service Providers, including Earthlink and others to provide them with the names and other data regarding the activities of the persons they wish to pursue in courts. The actions taken by these organizations are part of a rising awareness in the public, and in the creative community, that intellectual property rights can be aggressively protected. Even though you might be thinking, "I'm just a little guy and how's this affect me?" You need to be aware of your rights and how to protect them.

In the art industry, you will find there are those who will blatantly attempt to steal your images and market them as if they were their own. There are things you can do that we will explore, but first, let's establish some basics. What follows is taken directly from the U.S. Copyright Office Website, and can be found at: *www.copyright.gov.*

WHAT IS COPYRIGHT?

Copyright is a form of protection provided by the laws of the United States (title 17, U.S. Code) to the authors of "original works of authorship," including literary, dramatic, musical, artistic, and certain other intellectual works. This protection is available to both published and unpublished works. Section 106 of the 1976 Copyright Act generally gives the owner of copyright the exclusive right to do and to authorize others to do the following:

To reproduce the work in copies or phonorecords; To prepare derivative works based upon the work; To distribute copies or phonorecords of the work to the public by sale or other transfer of ownership, or by rental, lease, or lending; To perform the work publicly, in the case of literary, musical, dramatic, and choreographic works, pantomimes, and motion pictures and other audiovisual works; To display the copyrighted work publicly, in the case of literary, musical, dramatic, and choreographic works, pantomimes, and pictorial, graphic, or sculptural works, including the individual images of a motion picture or other audiovisual work; and in the case of sound recordings, to perform the work publicly by means of a digital audio transmission. In addition, certain authors of works of visual art have the rights of attribution and integrity as described in section 106A of the 1976 Copyright Act. For further information, request Circular 40, "Copyright Registration for Works of the Visual Arts."

The U.S. Copyright Website is the best place to begin to learn about how to best protect your own copyrights. It is urgently suggested you invest the money to copyright each piece of art you create so that gives you the strongest legal standing in the event you need to take action against a company or individual who is violating your copyright. It will only cost you $30 per piece to give yourself the best safeguards for your intellectual property rights. You can register using this link: *http://www.copyright.gov/register/visual.html*

There are compelling reasons why you should register your work with the copyright office. Most importantly, without registering, you can only sue for damages. Regardless, you are protected by law, but are limited to only forcing an offender to cease and desist from using your images. Take the time to educate yourself on this important subject. Your images are your lifeblood and your copyrights are their safeguard.

An outgrowth of political activism by and on behalf of artists began with efforts in the 1980s to regulate fair treatment for artists in the course of art trade. The most noticeable outcome of that movement was the Visual Artists Rights Act of 1990; title VI of the Judicial Improvements Act of 1990, Pub. L. No. 101-650, 104 Stat. 5089, 5128, enacted December 1, 1990. (VARA). You can learn more about it on the U.S. Copyright Office link above.

This act contains some of the strongest safeguards yet for artists. It was put to use recently when the artist Wyland, who is covered elsewhere in this book, took issue with General Motors. They wanted to paint over one of his whaling walls in order to put up an ad for the new Pontiac G6. An article that can be read at: *www.freep.com/news/locway/whale16e_20041116.htm,* in the Detroit Free Press reported an outcry from the public who did not want the artwork destroyed.

I know from a personal conversation with Wyland regarding his mural on the Long Beach Sports Complex that he would strenuously fight for his rights as an artist to keep that and all the other public murals he has painted from being destroyed or violated in some other way. He told me about a challenge he faced when the nearby Aquarium of the Pacific was attempting to have his mural removed because they thought it confused the public as their then new facility was about to open. They feared too many people would mistake the Sports Complex for the Aquarium since they are only a couple of blocks apart. The Aquarium backed off when he stood his ground and the mural remains on the building today.

You also should avail yourself of some of the better books on the subject of legal right for artists. Tad Crawford has written several books for artists including, *Legal Guide for the Visual Artist* and *Business and Legal Forms for Fine Artists,* both of which should be in your library.

Another book to consider is *Electronic Highway Robbery: An Artist's Guide to Copyrights in the Digital Era* by Mary E. Carter.

Daniel Grant in his excellent book, *The Business of Being An Artist*' gives the subject of artists and law extensive review and coverage. The volume is one of the better books that attempt to be comprehensive regarding the art business that I have in my library.

There is more to understanding and acting responsibly on your own behalf regarding copyrights than reading a few books or merely registering your images with the Copyright Office. If you already subscribe to *Art Business News*, *www.artbusinessnews.com*, you likely have seen articles by Joshua Kaufman, Esq., who is a partner at Venable, Baetjer, Howard and Civiletti LLP in Washington, D.C.

Attorney Kaufman provides great insight into legal matters as they pertain to the art business, and in particular with the art copyright business, in his regular columns there. If not, get busy and apply for your subscription immediately. He also is legal counsel for the Art Copyright Coaltion and the Art Publishers Association, which you will read about shortly.

You need to know if you give "someone" your original to photograph, scan or print, and "that someone" makes a derivative version of that original in the form of a photograph, digital image or other form, you do not own the copyright to that derivative unless "that someone" has signed an agreement with you acknowledging your ownership of that derivative version.

Sure, you trust your photographer or your printer, but if you have a monster image that gains tremendous sales and the derivative falls in the hands of an unscrupulous person, you have a serious problem on your hands. You need to protect yourself by having vendors sign releases of any copyright ownership for any work they may do on your images.

Another way to gain more knowledge is to take an informative tele-class from Ami S. Jaeger, J.D., M.A. in Sante Fe, N.M. through Artist Career Training *www.artistcareertraining.com.* Ami, as I am, is on the board with A.C.T. as an Art World Expert. You read more about Ami at: *www.legalart.biz.* Contact her or check out the A.C.T. Website to find out when her tele-classes are offered.

While you're on the site, review all the offerings. A.C.T. is a wonderful resource with an extensive array of classes and services for artists who are serious about improving the marketing and business aspects of their career. If you need mentoring and help keeping on track with what is important, A.C.T. is a terrific way to get assistance and to belong to a community of artists who also are diligently working towards improving their own careers.

I owe a debt of gratitude to Ray Granger for helping me learn more about the information that follows on the organizations discussed below. He is a long time industry figure and for many years has been a marketing and sales executive for the publishing arm of Joan Cawley Gallery in Tempe, Arizona. Ray is active with both the organizations you will read about now.

There are a couple of other things you can do besides boning up on copyright issues for your images that will help you be more professional and join in the fight against art copyright infringement. The first would be to join the Art Publishers Association, *http://apa.pmai.org.* This association has many benefits, including the opportunity to get to know some of your competitors.

The APA is established under the PMA (Photo Marketing Association), but don't take it to think they are more interested in photography than art because they are not. The PMA gave them an independent way to form and affiliate with an organization that is not owned or tied to any large art industry publisher or tradeshows.

What it offers is more than a chance to socialize with other publishers. Its membership list reads like a Who's Who of Art Publishing. But don't be intimidated by that. In fact, welcome the opportunity to meet some of the industry leaders and get to see what makes them tick.

Some of APA's goals as stated on its Website include:

- Growing the market for art
- Providing a forum for open communication and networking
- Promoting consensus on publishing terminology and standards
- Educating retailers and consumers
- Promoting high standards of business conduct including copyright matters
- Providing market research and industry information
- Managing our businesses better

I believe you can make up any expenditure and effort from being involved with this organization simply by paying attention to, and learning from, the activities of the Uniform Standards Committee. The goal of the Committee is to help educate artists, publishers and galleries about art prints and the language used to describe them. They are working to devise a cohesive vocabulary aimed at eliminating confusion in the industry and ultimately with consumers.

The goal is to eventually have publishers, artists, and art dealers agree on established standards of language and descriptions of various art media so everyone involved will be correctly educate consumers about limited editions and to ensure they are properly represented.

The committee recommends all publishers provide each consumer who purchases a limited edition work with a certificate of authenticity that accurately and fully describes the artwork. With the assistance of its attorney, Joshua Kaufman, the Uniform Standards Committee has produced a booklet to help publishers design and write certificates.

The committee also has established a process by which an artist or publisher can have a certificate of authenticity approved by the APA. By providing a stamp of approval, it will further strengthen the confidence of consumers. Full disclosure is the most significant part of its mission to build consumer confidence.

A variety of certificates of authenticity will continue to be used because establishing one that works for everyone all the time has proven to be difficult. Laws vary from state to state and require different terminology regarding disclosure. Currently, some 14 states have different laws with regard to COAs (Certificate of Authenticity). It would cost you more in legal fees and your own time in research to learn how to comply with these varying laws than the cost of your APA membership.

You know, or should know, that a COA is necessary for any kind of limited edition art. Not only is it the law in many states, it shows your professionalism and the respect you have for your art and your customers. Some pundits in the industry believe, and I tend to agree with them, that COAs should be offered with originals. By including a COA with your original, you also can use it as a chance to reiterate your copyright on it. That is, although you have sold the original, you maintain the copyright for the image and as such maintain the sole right to reproduce or publish the artwork or any facsimile of it in any medium. Face it; the average collector buying an original is clueless as to the ownership of the reproduction rights of the piece.

There is another organization that works with copyrights and that was organized to fight copyright infringement in the art industry — one in which you also should consider membership, if not at first if your budget is tight, then have it earmarked for future participation. It is worthy of that at least. I'm speaking of the Art Copyright Coalition. It was formed by many of the same members of the APA after years of seeing their images blatantly knocked off, but by no means exclusively, by companies with foreign ownership that have U.S. operations.

In my 15 years with *DECOR* and Decor Expo, some of the most vitriolic exchanges (and there were some good ones) I witnessed among exhibitors, and even with the management of the various tradeshows, was over the question of allowing certain companies to exhibit. Most often, it involved the legal status of companies who sold inexpensive assembly-line-produced, imported oil canvases.

While tradeshow management explained they could understand the frustration these exhibitors felt, they contended they could not restrain a company from an equal opportunity to exercise free trade without some legal standing to keep them from exhibiting. In other words, produce an order to cease and desist or some other legal means and then we can get something done.

The ACC has a Website worth viewing. Spend some time there to get a feel for the work they are doing:

www.artcc.org. You will begin to understand why it will be important for you to belong when you see the list of benefits the organization provides for its members. Here are some of the membership benefits as outlined on the Art Copyright Coalition Website:

- Templates for cease and desist letters for independent use.

- Initial review, analysis and discussion of infringement reports rendered on full member's behalf, up to 75% of dues paid.

- Be notified of all infringement investigations where the ACC has pictures of infringements, or suspect that member's imagery may be infringed upon.

- Be allowed to be a party (in a JET) to any litigation filed against a suspected infringer (when the member's rights are infringed) by a group of ACC members.

- Be able to freely exchange information with other members about infringers, or tactics used to detect infringers, and to get legal results against them.

- Be allowed to pass suspected cases of infringement to the ACC, who will investigate the matter to determine if an infringement has happened, and to try to find out if other members have been infringed as well. If several members' rights have been infringed, the ACC will, through its counsel, offer legal advice to the group of infringed publishers, and encourage them to take joint action, sharing in the costs. This benefit is only for members who have a balance on their account (as noted above).

- Receive the ACC's booklet, "Basics of Self Investigation of Copyright Infringements," which guides the member in collecting the first stage of information, setting the best possible stage for a deeper investigation or legal action.

- Receive a data feed of trade show surveillance, if possible before the tradeshow ends, so that members can ascertain if they are being infringed at the trade show, and thus strive to respond with legal action (such as seizures, trade show expulsion) before the tradeshow closes.

- Access to the secure section of the ACC's website.

- Benefit from publicity and education of the public by ACC against copyright infringement/ counterfeiting.

- In the future, benefit from ACC's development of a coordinated program to work with Customs officials to prevent unauthorized importation of infringing works into members' home countries.

Who are you going to believe, your lying eyes, or me?

– Groucho Marx

Chapter Nine

Finding And Working With A Publisher

> **True art is characterized by an irresistible urge in the creative artist.**
>
> – Albert Einstein

From experience, I know many of you, well before you have even read this chapter, had already decided the road of the self-publisher was not for you. No matter how you got here, if working with a publisher is your choice; I applaud you for having made the decision. For most artists it is the easiest, best, and safest and sometimes only course of action.

It doesn't matter that even if you yearn a little to someday own your own publishing business but can't swing it right now, the information you'll find in this chapter will help set you on the way towards success with an established publisher.

Although getting started with a publisher for some of you will be a step towards eventually publishing your own images, for now you can be happy to have someone else take care of the myriad of details that are involved in publishing art. Some will find it's just easier to let their publishers handle the details and they will be right.

While working with a publisher is far less expensive and complicated than publishing your own work, you still need to work hard at first finding the right publisher and maintaining the relationship. Sounds like a marriage, doesn't it? There are definite similarities, and as with a marriage, the more you work at it and keep communication open and honest, the better the result will be.

In Chapter Six, I offered bullet points to indicate what I believe are both the major merits and pitfalls for artists who are contemplating self-publishing. Here are the major merits and drawbacks to working with a publisher.

Merits of Working With A Publisher

- You get feedback and advice on what to create
- You can concentrate on creating art
- You get recognition and sales through expensive marketing efforts
- You get access to the dealer network
- You get access to the licensing contacts and established business relationships

Pitfalls of Working With A Publisher

- You get feedback and advice on what to create
- You make much less money, typically 8-10% on the wholesale price
- You are in a stable of anywhere from a few to dozens of artists
- Your publisher has established stars that pay the bills, and if you aren't one of them, you might have to struggle for recognition

Of course, that you and others might take issue with these points is the nature of a good debate among intelligent people who have a difference of opinion. Unlike a former tradeshow director with whom I once worked, I welcome questions and debate. That director had a sign on his desk that neatly summed up his opinion about negotiating and debate, it read: "Be reasonable, do it my way!"

There are countless stories of artists who work in concert with publishers that make sizable incomes from the revenues those publishers generate for them. To avoid misleading you, it needs to be mentioned there also are countless others for whom revenue from publishers is a mere steady trickle. I wish I could accurately account for the difference between the stars and those back in the pack.

If pressed I would again call upon my always fallible S.W.A.G. Factor, and say most successful artists who work with publishers instead of self-publishing have in common attributes or traits similar to those of the self-published artists we discussed in Chapter Four. To refresh your memory, successful self-published artists have these things in common:

- Talent
- Art that resonates with a large group of collectors
- Financing
- Personnel, usually in the form of a spouse, devoted family member or close friend
- Willingness to prodigiously continue to create art that is in the same thematic range in order to continue to supply the dealer and collector base
- Ambition

If you look over the list, having personnel and financing are items not as critical to success when working with publishers. Those two factors are at the heart of why many otherwise capable artists end up working with a publisher.

Beyond that, the rest comes down to believing in yourself and working hard at all the aspects of the business you can control.

To some extent, as with many other facets of publishing, success in working with a publisher comes down to talent, luck, ambition and timing. Not all of those are within your control. You can't control that you will encounter someone who will make a huge difference in your career or like your work, although being ambitious surely can help in that regard.

When I think of artists in other fields, I sometimes wonder how they would do in other eras. For instance, look at Bob Dylan. Perhaps I'm reminded of him because I recently read his fascinating bestselling autobiography "*Chronicles, Vol. 1*" which should be must-reading for any fan or anyone interested in how folk music helped to shape the culture of the baby boom generation in the second half of the last century. Dylan was a scrawny 19-year-old kid from Minnesota when John Hammond plucked him from obscurity and changed his fate.

I have the utmost respect for Bob Dylan. His contributions to rock and folk music and his towering influence over pop culture and other recording artists in the 20th Century seems almost as unreal as his lyrics sometimes. Still, I have a hard time believing if he came onto the scene today that is dominated by hip hop and slick pop that he would have found the same success.

Destiny and kismet are always at play and still you can only strive but to do your best. To some degree, you have to have the right look at the right time and get connected to the right people — as in publishers — to get to the top of the heap in art print publishing.

That doesn't mean you will suffer in despair somewhere further down the ladder from the top of the heap. I suppose with his enormous talent that Bob Dylan would have carved out some kind of rewarding professional career as a musician and songwriter or lyricist if he were starting out today. At one point in his book, he makes a comment that he should have been a manager. With his drive and belief in himself, he would have been great at it had he taken that path.

It all depends on how you play it and how you manage your business and the opportunities that present themselves to you. You can enjoy a nice living as a print artist if you apply yourself to being as successful as you can by keeping as many of those above listed bulleted traits working in your favor.

Deciding to work with a publisher being the right route for you to take was your first decision. Following along right behind it in importance is deciding in what medium you wish to see your images reproduced. So, what format best suits your images? Sometimes this requires serious contemplation on your part.

Sometimes the content or size will help guide the decision. Sometimes knowing your work relates to a certain theme or style will help dictate format and publisher choices. For instance, if you are a wildlife artist, you are probably looking at offset limited edition prints and giclées. That is where the bulk of the market is found. For starters, you'd want to look for publishers like Mill

Pond Press, Wild Wings, Hadley House or Greenwich Workshop, among others.

You need to be as proactive as possible in making decisions; it is your career after all. Yes, you are putting your career in the hands of a publisher, but you have to maintain as much control as possible. That means deciding beforehand in what medium you want to see your art reproduced. Ultimately, you may not be able to dictate this choice and there may be cases where you are asked to reconsider your decision. You can decide at the time what is best for you and try to be as flexible as possible, but forming your own strong opinion honed by your instincts and investigation should negate much controversy long before you begin contacting publishers.

This sounds like a no-brainer, but I'll say it anyway. If you know your images are really best suited for a poster catalog, you should focus on those publishers who primarily or exclusively publish posters. You are much more likely to find poster publishers advertising and exhibiting to the *DECOR* market.

If you have determined your images belong in the limited edition market, which can mean anything up to giclées and serigraphs, you should narrow your search to those publishers whose emphasis is on that end of the market. At one time, a division existed in the limited market between offset lithographic limited edition print publishers and giclée publishers, but it faded away.

You will still find some publishers who publish exclusively in serigraphs, or in serigraphs and some giclées and whose price points put them at the top of the middle tier of the industry. A tour of the center most booths in the "Platinum" section of ArtExpo is a quick easy way to determine many of the players in the high rent district of limited edition publishing.

The "Platinum" section emanates out from the center of the show much as you'd find the center of a town at the town square. It's the place where all the major avenues in the show intersect. It's where most of the companies with the biggest budgets vie with each other for space. It's where many of the industry's highest profile publishers strive to be.

I've encouraged you elsewhere in this book to attend ArtExpo and here I am doing so again. If it is repetitive, it is because I believe there is no better place for you to get smarter fast about your future in the business than by immersing yourself in the experience of ArtExpo. It's that important to those with the print market in their future.

However, if posters are where you are headed with your print career, then the Decor Expo Atlanta show should be the first place you want to go, followed by the Decor Expo New York show.

Both Decor Expo Atlanta and New York shows now run concurrent in the same facilities as ArtExpo, which is an enormous convenience for the whole industry, so when you go you need to make time to absorb both of them. Underestimating how long you will need to even walk these shows, much less take time to try to meet people is a mistake many people make.

If you can spend three days at these shows, you should do it and no less. Of course, if you don't have the financial budget or time to allow the luxury of three days, you'll have to work smarter to make sure you see what is important to you for your personal growth and knowledge base needs.

If you can't get to New York or Atlanta, don't stress over it. Sometimes things are not possible even though we know they ought to be. Keep in mind there are plenty of successful artists, especially those repped by poster publishers, who have never set foot in any tradeshow. It's not a do or die criteria to the advancement of your career. You can find your path to publishing success in some other ways. You just have to figure how you are going to learn about publishers without seeing them all together under the same roof.

If you haven't had the chance to peruse the Websites of some of the better-known publishers in the industry, it should be on your **must do** list. Another way to do that is to find a local gallery that carries works of artists you admire and then spend some time there looking through its catalogs, but with the following caveat. Having spent time working in a gallery, I can tell you many owners find the visits by artists who use their galleries for personal field trips much as a museum to be a nuisance.

Keep in mind that galleries are there to make money, not entertain artists who aren't really prospects for them. So, be respectful of the time and space, and don't put on ruses as a buyer so you get personal attention from a salesperson. Politely ask if you can quietly review the catalog in question.

If you are visiting poster galleries, they typically are accustomed to having people take time to go through their books. But, that doesn't mean you shouldn't have the same respect for them as you do original or limited edition galleries. They all need to eat and someday one of them might have your work in its shop. Imagine that you wouldn't want the owner or salesperson distracted by an artist with no buying intention when a customer came in who would be a great prospect for your work.

I cringe at making these suggestions, but since I also realize there are few ways to get your hands on a catalog, I do it with the admonishment to be very respectful of the gallery in the process. It's an unfortunate fact that because catalogs are so expensive to produce, many publishers will only send their catalogs to qualified dealers. Most sell catalogs only to dealers or require a substantial minimum order before they provide them at no charge.

Many publishers will not entertain any dialogue with collectors or consumers and typically, they redirect them to the most convenient gallery that carries their work. They know the quickest way to gain the enmity of their coveted dealer base is to begin to compete with them. Still, some do have distribution schemes that create conflict between their dealers and their direct sales to consumers. Thomas Kinkade has many company-owned galleries, for instance. Channel conflict is not the norm in the art-publishing arena, as most publishers prefer to concentrate on their work as wholesalers in support of their dealer base.

Fortunately, the Internet makes everything easier these days, especially for sleuthing to gain intelligence on companies that you want to work with. While viewing a high quality four-color rendition of a print or poster is much richer than the maximum 72 dpi images you see online, you can still gain a good understanding of the offerings from various publishers by viewing their Websites.

Publishers are in the business of finding new artists and new images. Go to any tradeshow and listen as buyers come into their booths. The first question is usually, "What's New?" It doesn't matter if the old stuff is still the greatest thing since sliced bread and is carrying the company and its dealers with 80% of the sales; everybody wants to see the new stuff. Publishers realize a serious part of their job is to be ready to look at new work and almost all have formal policies and methods of reviewing the works of new artists.

Publishers also have criteria for what they would consider the ideal artist. What follows is a composite of talking with many publishers over the years to ascertain what qualities they seek in an artist. Here they are:

Publisher's Wish List

- Work has commercial appeal
- Work is consistent
- Easy to work with
- Works on deadlines
- Is coachable on specific design requests
- Is dependable and reliable
- Is available and easy to contact
- Is flexible and versatile

More on Contracts

There are two basic ways publishers pay artists, either a royalty or a flat fee. Royalties take much longer as money comes only as inventory is moved. Most publishers are going to want to pay you in royalties; it's to their advantage. You can always attempt to negotiate an advance against royalties. If you are successful in getting the advance, try to make it non-refundable. Perhaps you use the second point to give away in a negotiation to get the advance. Sharpen your negotiation skills. They really will come in handy more often than you imagine.

You need to get into the contract the specifics of your deal on items large and small. Who pays for shipping and insurance of your original to and from the publisher? If you have provided other valuable materials, such as 4" x 5" transparencies, framing, et cetera, make sure your contract states they will be returned to you and when.

Proofs that are produced without the intention of being sold are sometimes called H/Cs, for the French term, *Hors d'Commerce*. This is used more often in "elite printmaking" such as etchings and serigraphs. It means before the sale. H/Cs are used for proofing purposes and sometimes in promotion to help create interest in the finished prints. All proofs and prints need to be accounted for in number and disposition (How they were used.)

How H/Cs will be used and disposed should be clear in your contract. This is especially true in the case of limited editions that are more expensive than posters.

Sometimes, H/Cs are denoted as P/P, which stands for Printer's Proofs. This is more common in offset and giclée printing. Your contract should state that all such proofs not used or destroyed should become your property and who pays insurance and shipping costs to get them to you. Likewise, if printing plates from for example etchings or serigraphs, are used in the production of the prints. They need to be canceled and saved with proof provided the artist or the actual canceled plates provided to the artist.

Do those Publishers' Wish List attributes accurately describe you? If you can honestly say yes to all of them, congratulations! Your success is all but signed, sealed and delivered. Reality is that for a multitude of reasons, few artists, can live up to every item on the list above. Publishers are aware that a wish list and reality rarely coincide and accept what they find in the best and worst qualities of artists with whom they work.

If you already have tried approaching publishers, you probably have received a variety of responses. As with life in general, it takes all kinds, and that means you will run into publishers who are friendly and gracious and those who are less than that. Don't be put off by anyone who is not receptive or friendly, your goal is not to make friends. Your first goal is to get an audience and your second is to be accepted by one of your top choices.

One thing all the best publishers do have in common is an established means of how they would want to receive artist submissions. Many conveniently have their submission standards posted on their Websites. A typical clear cut example of how publishers offer submission information can be seen on the Website of one of the industry's top poster publishers and distributors, Image Conscious: *www.imageconscious.com/a/submissions.html.*

In this case, the staff at Image Conscious provides you with everything but how to paint the picture. Follow the explicit guidelines and hope for the best. They have made it easy for you to submit your work to them by telling you in advance all the details on how they like you to submit your artwork.

If you have questions that are not covered on a publisher's Website or in the submission materials, don't be shy. Call and ask for clarification. That could include how you can follow up on your submission. There is nothing wrong with following up. It can be construed as professional if done in a polite, even persistent manner. You can, on the contrary, also be viewed as highly unprofessional if you just become a relentless pest.

If you let your eagerness or anxiety push you over the line into the pest category, you are damaging your chances of succeeding in getting your work published.

Remember to take a few deep breaths and practice what you want to say before you make those follow-up calls.

Obviously, if you follow the guidelines publishers provide for you, you are going to have a better chance of getting a favorable reception. Always be courteous and remember that following the rules won't help if what you paint is not what they are looking for. If you have done your proper analysis before submitting, you will cut down on both the number of places you submit and the number of rejections you receive. Not all publishers have clear submission information on their Websites as Image Conscious. With some, you will have to contact them to have guidelines sent to you.

Don't risk everything on one endeavor. Don't submit to one publisher because you admire some artist in the line, or for whatever reason. You should submit to several publishers at a time to give yourself greater odds of having success. It can take weeks, if not months sometimes, to get replies from publishers. Some publishers review for new artists only at certain times of the year.

In some cases, they may tell you that you are in the running, but there will be more decisions and things can drag out for agonizingly long periods. This is the nature of the business to be aware of before you get started.

With regard to submission standards, whatever you find out from a publisher that you have targeted, pay close attention to them and follow the instructions. How you respond can show the publisher you are both eager and professional. Giving a hint that you are someone who would be easy to work with is always in your favor.

Being a prima donna will only work for you if you are in a position to deliver images that are going to be instant top sellers in the market. Otherwise, you'll find that being difficult will end up getting you less attention. This is especially true where a publisher is putting out a thick catalog with potentially hundreds of other artists represented in it. Their perspective tells them there are always options for coming up with a similar look but without the hassle.

How Do I Find Art Publishers?

This is the easiest part of the whole process. That's because there are a finite number of players in the industry to they are almost all visible to a degree. If you haven't subscribed to the industry trade publications, I reiterate that you need to make that another item on your **must do** list immediately. The trade pubs are your windows to what is going on in the business. This is even more critical to you if you aren't finding a way to get to the major tradeshows in March and September.

Because it is so important this thought bears repeating; in order to succeed in this business you need to be a student of it. You need to learn how things are done, and who and why they are doing them. Trade magazines are full of this information. Not just the articles, those are aimed at the primary readership of dealers with ideas and suggestion on how to operate their retail businesses. The editorial columns and advertising are full of useful information for you to glean about the activities of publishers and artists.

To start with, many of the top publishers are frequent, if not monthly advertisers, in the trade publications. The primary trade publications for artists to be reading are *DECOR, Art Business News* and *Art World News*. The latter is an independent publication while the first two are part of the International Art and Framing Group that is a subsidiary of Pfingsten Publishing LLC. You will find more information about trade publications listed in the Resources section at the end of this book.

Once you start to read and carefully review the content of these publications, you will begin to understand which publishers are most likely to be the best candidates for your images. You'll see some that publish a variety of artists, and others that specialize in one or two. But as discussed earlier, not every publishing company follows my prescription for success by using a combination of trade advertising, tradeshows, and publicity and so forth. That means there are good candidates for you to be submitting your work to that aren't regularly featured in the advertising pages of the trade magazines — some not even on an infrequent basis.

You will find some of those who are missing from the magazine advertising pages in a few other places. The Internet is one place you can look, but it could be tedious going. I ran the search term "Art Publisher" through Google and it came back with a mere 18,800,000 results. That's a lot of digging. Still, if are patient enough to wade through the first 50-100 pages, you'll probably uncover some companies you wouldn't find otherwise by sticking to trade magazines and tradeshow directories. Overall, searching the Internet for publishers to approach is probably not a good use of your time, unless you have art that falls well out of the mainstream.

By now, you know my biases. I believe in using all the tools available to get a product to market. It is hard for me to understand why you would want to eliminate from your plan opportunities to get your message in front of your best prospects. My advice is to try to hook up with the more prominent publishers because they normally are the ones using all the tools.

Here are the best places to find the industry's top and most visible publishers:

- Trade Shows
    - ArtExpo New York
    - Art & Framing Showplace
    - Decor Expo New York
    - Decor Expo Atlanta
    - ArtExpo Atlanta
    - West Coast Art & Frame Show
- Trade Magazines
    - Art Business News
    - Art World News
    - DECOR
    - DECOR Sources
    - Art Book/Frame Book
- DECOR Sources Online
- National Art Consumer Publications with art print interest
    - Southwest Art
    - Wildlife Art News

Assuming your work has commercial appeal, you should be able to find a publisher for it by pouring through the above resources. I've mentioned attending tradeshows as a way to jump-start your art print marketing education. A valuable item you'll come away with from any of these shows is the show directory.

The show directory conveniently tells you the name of every company exhibiting at the show and gives all their contact information. However, most will not include the name of someone to contact, but everything else is there for you to begin building your own publisher database. If you aren't able to make it to the show, contact the show producer as soon as it is over to inquire about purchasing a copy of the show directory, as their remaining supplies of these directories will be limited.

By subscribing to and reading the trade magazines every month and noting the various publishers who advertise in them, you will begin to have a general idea who some of the best prospects are to approach. The same holds true for the tradeshow directories. These are invaluable sources for you to research for publishing companies to contact.

Both *Art Business News* and *DECOR* publish annual directories and these can help you find those publishers who aren't the most visible, meaning they advertise less or not at all and exhibit at fewer shows. Within the pages of these directories, you can find almost all the remaining publishers worth contacting. As with the tradeshow directories, you'll be able to cross-reference the publishers with the listings of companies, including all their contact information.

Decor Sources also publishes a database online *www.decor-sources.com* that is searchable. This is yet another way to dig for publishers, although in this case, it is a duplication of their annual directory issue. It will be a question of which is easier for you to work with.

Only two consumer magazines are on the list. There are many more, but few if any would have advertising or participation from the kind of publishers you seek. While they might have editorial you find compelling, you aren't going to discover many prospects for publishing companies to approach in the pages of *Art News*, *Art in America* and so forth.

When approaching the top echelon of publishers, you'll want to keep in mind you will be competing with lots of other artists for visibility, for space in their marketing materials. For instance, once accepted, your images might only be shown once a year, if at all, in their trade advertising. But that doesn't mean members of their sales staff aren't hustling your images to appropriate buyers behind the scenes. For many publishers at this level, trade advertising compares to window dressing for a department store. It highlights the new fashions, but the real shopping only begins when a buyer enters the store, or in your case, opens the catalog.

For poster publishers with hundreds of images, they rely on their sales staffs and their marketing staffs to send samples of new suitable works to their best customers and prospects. Even though the mom and pop retailers might not see your image in the latest ad from these less high profile publishers, many of the volume buyers and top dealers will be sampled with your work as soon as they are available in proof format.

Publishers know what they are doing even though many new artists really have to wonder if they are getting any exposure at all from their relationship. In some cases, things won't work out and you'll feel the need to move on once your contract is over. Unless publishers know they have a certified winner with some image or line of images on their hands, they know they need to work on building awareness for you and your style of art.

It can take a while for your images to break through and begin to establish a following initially with the dealer base and with the buying public. Patience in these situations is required, but understandably, it can be tough. This is where having a professional and courteous relationship with your publishing company personnel will work in your favor. Developing and maintaining these relationships is the best way to be kept abreast of your circumstances.

Publishers typically pay 8-10% (sometimes higher for top producing artists) of the net wholesale price they receive for a print. That means if they sell your image at a $40 retail price with the typical 50-50-20% discount I discussed in Chapter Five, they end up with $8.00 and your take is $0.80 (10% commission) - $0.64 (8% commission) per piece.

Seeing the economics broken down to that fine point is demoralizing to read, I'm sure. But, take heart because plenty of artists working with publishers are doing nicely when you consider they are also licensing the images through them as well. We'll discuss licensing in a later chapter. For now, keep in mind it can be an important source of income for you and when you layer it with the publishing royalties, it will be lucrative.

If you can see it is not going to be enough, you ought to think about getting your originals into galleries and finding other ways to market your work. Some publishers require an exclusive with you, meaning they don't allow you to shop the work they decline with other publishers. Some don't make this distinction. Remember my advice on learning some negotiation skills? You can see from this instance alone why it would be good to have worked on them before starting to seek a publisher.

To help augment your income, you may have to self-publish some of your work in giclée or some other format and seek local galleries or exhibit at some local or regional shows to bring in more income. This is where having an entrepreneurial streak can help. On the other hand, if you are in the situation where you don't have to pay all the bills from your art, you can be patient and let things develop through working with your publisher.

You also could realize you have dual careers going at once: one aimed at the fine art originals gallery or art show market and another aimed at the poster and print market. The aforementioned story about Arthur Secunda is an example, as is that of Richard Thompson. They are just two of many that have accomplished this feat. Another approach is that some poster publishers have artists painting for them under an assumed name.

While personality is important in the marketing of a limited edition artist as with Wyland, Steve Hanks, Thomas Kinkade or P. Buckley Moss, it is really far less important to publishers with as many as several hundred artists whose work they represent. Most poster publishers don't have their artists making personal appearances for them. You can retain anonymity from the public with a poster publisher.

Having a theme attached to your work will help you in the marketing of it. When working with a publisher, once you are established, it doesn't hurt to let the publisher know you have the capability and desire to paint in other themes. This is especially true if you are willing to take direction and be coached by the art development team. If you can help them fill a need in a product line without having to discover yet another new artist, you make yourself that much more valuable. This multiple hat scenario isn't going to work out for every artist, but it's something to keep in mind as you develop your relationship with a publisher.

Depending on your needs, you could consider working with less prominent publishers on what could be called the second level. Working with a less visible company doesn't necessarily mean you will make less money with them. Some of them are connected to the contract design market. Contract design is the commercial side of interior design. It involved those interior designers who work on commercial buildings including hotels, office buildings and hospitality operations. They need and order vast amounts of artwork at one time. Others might have direct connections with volume framers and are bypassing the retail section of the market. Still others have found a profitable niche that allows them to operate without trying to compete head-to-head with larger publishers.

Finding these less visible publishers is more difficult. They do surface at the larger tradeshows and in directory listings. If you are using the publicity columns of the trade publications, advertising or exhibiting yourself, they will find you, too. Typically, they have smaller booths at tradeshows, use less advertising and have an overall lower profile than industry leaders. It's easy to figure out who is who through mere observation or learning how many artists or images they publish. Don't forget, it's the job of every publisher regardless of how large or small to look for new artists.

We discussed that this book is aimed at the middle and lower tiers of art being sold. Often, it will be tougher to crack middle tier publishers' lineups than those open edition and poster publishers on the bottom tier because they work with fewer artists. That means when bringing on new talent they can be fussy in reviewing for it. You will find these publishers advertising primarily in ABN and AWN and exhibiting at ArtExpo.

In some cases, having gained a reputation for yourself in the print industry or art arena will be helpful as often it will lead to relationships with people who are in a position to introduce or recommend you to someone. Over the years working with *DECOR,* I made connections for artists with publishers.

Another way to gain visibility with potential publishers is to exhibit at shows where publishers might be on the lookout for new talent. One show is the SOLO show, which is part of ArtExpo. For some artists, this is a great opportunity to gain exposure to the market and, in particular, to publishers who will be attending ArtExpo anyway. Below is the information on the SOLO show taken directly from the International ArtExpo Website:

SOLO Package:

International Artexpo makes it easy for individual artists to exhibit their work with the following cost-effective package specifically designed to meet an artist's needs.

Exhibit Package
10' x 4' space

Exhibit Features

- 10 foot high fabric-covered hard walls
- Two halogen lights per 10' x 4'
- One chair per 10' x 4'

- Fully carpeted exhibit hall
- Trade and/or consumer VIP Passes for top clients
- Booth signage
- Booth cleaning
- Complimentary drayage and storage of empty containers

Specifications

- Open to artists only (no galleries)
- Only original artwork can be displayed by artist
- Only one artist allowed per booth
- No more than 2 booths per artist
- All artwork must be hung subject to Show Management approval
- No tables allowed in booth
- Artist must submit slides or images with application

Do you recall the *Time* magazine article I referred to in Chapter Seven that derisively referred to populist artists who were noted in the piece as palette-to-paycheck artists? That article noted 30 artists were millionaires or multi-millionaires. If you thought they were all self-published artists, you would be wrong. Some of them work with established publishers who do the printing, marketing and sales of their work.

Here are a few examples of high profile, high-income artists that do or have worked with publishers:

- Bev Doolittle–*www.greenwichworkshop.com*
- Robert Bateman –*www.millpondpress.com*
- Steve Hanks – *www.hadleyhouse.com*
- Thomas Pradzynski – *www.palatino.net*
- Royo – *www.royoart.com*
- Carl Brenders- *www.millpondpress.com*
- Charles Wysocki –*www.hadleyhouse.com*
- Warren Kimble – *www.wildapple.com*
- Bob Byerley – *www.wildwings.com*
- G. Harvey – *www.somersethouse.com*

Access to information on what the actual income is for any of the above artists is not available. But I can say with confidence each one is doing well from the financial relationship that they have with their publishers and most assuredly, some are millionaires or multi-millionaires through those relationships.

Most on the list are working with publishers who specialize in limited editions, as opposed to posters and prints. Nevertheless, there is variety in that some of the publishers listed have large stables of artists, while others are single artist publishers as Royo at Triad Publishing, or represent just a couple of artists as with Pradzynski and Manuel Anoro at Palatino Editions.

There are cases where artists working for poster publishers earn large incomes, especially when you consider the sale of their images and the royalties that flow from the licensing of their images. Licensing has become an important aspect of art publishing revenues and has flourished into a full-fledged portion of the operation of many art publishers. This subject is important enough that it will be discussed in more detail in a later chapter.

Let's assume you've gone through the steps of targeting publishers, contacting them and have found one that wants to publish your images. Congratulations! You've taken a giant step towards making your art career full-time and prosperous.

This is a timely place for this quote:

> A verbal contract is not worth the paper it is printed on.
>
> – Louis B. Mayer

When it comes to dealing with a publisher, you'll find any reputable one will have legal documents that protects it and you and binds the two of you together for a period. Take the admonishment in the quote above seriously; <u>don't deal with anyone on a handshake.</u> Sure, there are stories of people who have had fabulous and lucrative relationships based on nothing more than a promise, but when it comes to your career, you can't afford to take that chance. My advice is to find a good lawyer who has experience in contract law and who insists on getting EVERYTHING in writing.

If you are scraping by on a starving artist budget, an alternative often suggested is to seek legal help through pre-paid legal programs. They offer legal advice at rates far below a typical lawyer. Most advice they dispense can be deemed reliable for simple transactions as wills; some divorces, creating a corporation, and so forth. However, I do not recommend that you settle for this kind of legal advice when it comes to reviewing a potentially complex contract that affects your future and your livelihood.

A good lawyer will not tell you whether to sign a contract. They usually take the stance that their job is not to tell you whether to skate on the pond, but rather how hot the blades, how thin the ice, and how cold the water is. It is left to you to make your own informed decisions.

The good lawyers also will admit that if their clients listened to every detail of possible negative outcomes from engaging in a contract, there would be no contracts, just stalemates with all parties unable to agree. That means after you have heard everything your lawyer has told you, you have to weigh the potential positive outcome against the negative and engage in good faith bargaining with your publisher.

Here again, as I have advised several times throughout this book, understanding the art of negotiation and mastering some basic skills will lead to more favorable results for you. Obviously, seasoned publishers dealing with an artist ready to sign a first publishing and licensing contract have a huge advantage. If you come into the negotiations with demands the publisher views as unreasonable, they can assume the position of "take it or leave it."

If you only ask for what you think you might get, you are likely to be scaled back from what you wanted. While if you ask for more than what you truly expect, you might get more than you estimate you were going to get. There you have Lesson One in negotiation techniques. Imagine your improved results if you study and polish your skills for such a scenario.

Even as a newbie, you always have rights and you should do your best to make sure you keep as much control over the negotiation process as possible. I mentioned before that being a prima donna is not the best way to get good cooperation with the staff of your publisher. That doesn't mean you can't insist on maintaining the integrity of the art you are submitting or have had approved for publishing.

For instance, if they would want to add graphic elements or type, or pump the red, or whatever the issue, you should maintain the final say on whether or not that will be allowed. Keep in mind they are operating a business that is fashion oriented and often their suggestions for changes are because they believe those changes will lead to more sales. So, if it is your masterpiece, you might want to refuse to allow it to be altered. On the other hand, if it is something you don't mind seeing altered, you are probably going to enjoy seeing more money in the form of commissions by being agreeable. At the end of the day, it is your choice and that's how it should be stated in your contract.

Since you are being paid on the sales of your images, you also should have some right to review that portion of the publisher's books that contain the record of the sales of your prints. That is not to say you should feel the need to audit your publisher's books annually, but merely by having the contractual right, you help keep the playing field more level.

To summarize those things to consider when entering into an agreement with a publisher:

- Pre-register your copyright at: *www.copyright.gov*

- Get it in plain writing that the artist holds the copyright

- State the number of copies the publisher is authorized to sell
- Set a time limit on how long the publisher can advertise and sell works
- Maintain some control over reproductions whenever possible
- Ask for conditions whereby your contract can be broken (contracts are made for divorces, not marriages)

Expect to have the images you and the publisher agree to license to be tied up for a minimum of three years. If publishers are going to take a chance on you, they need to know they have enough time to earn a profit from the deal.

Anything else you can get into the deal to help sweeten it for you is worth asking for, or at least use as a bargaining chip. If you have done your homework, you probably already know at least some of the means your potential publisher uses to market images. You'll want to ask:

- If they have a catalog and how often do they update it with supplements
- Where and how often they advertise
- What shows they exhibit at
- Do they have a licensing division
- What portion of their business is with the volume production market and how often do they sample them
- What do they do with other marketing tools such as direct mail and fax blasts, Website and email marketing and so forth

- What is their plan for breaking new artists

Depending on the nature of your relationship to start with, the openness of your publisher to share details and probably, to some extent, how eager they are to publish you will decide how much of the above list they will be willing to share with you (Mr. or Ms. New Artist on their roster). Don't consider it your inalienable right to be able to demand answers to these questions, but it doesn't hurt for them to know you are interested in how their marketing process works and how it will affect the sales of your images.

It's highly unlikely a publisher will commit to you when your images will appear in any given issue of a trade publication, or be displayed as a framed print in their booth. However, they ought to be able to give you some assurances and, in some cases, even concrete details how they intend to market your images.

I'm certain all publishers would love to assure all their new artists they are going to experience unqualified success with their imagery. Unfortunately, it's not that easy. Publishing is a business that is fashion and trend oriented, and as such, you or they can never be certain what is going to work. With their experience, they often have good judgment about what prints will be hot, but they are often surprised when one from the line gets legs.

Having a print or poster get legs, or an artist that takes off with a whole series of images that fly off the shelves and ends up with endless licensing deals, are what makes the publishing business profitable. Those winners pay for lots of dogs that won't hunt. What you hope and pray for is your art turns up in the winner's category.

The facts are it's nearly impossible in advance to predict what is going to happen once art is created and ready to be marketed and sold. This is true in other areas of the arts. Consider this quote from a February 23, 1993 article by Bruce Haring on *www.Variety.com*

> Eric Clapton, the legendary rock guitarist/singer long deified by his fans, ascended to Grammy heights, capturing six prizes, including album, song and record of the year, to highlight the 35th annual Grammy Awards last night at the Shrine Auditorium.
>
> Clapton, whose single "Tears in Heaven" detailed his heartbreak after his young's son's accidental death, admitted when he accepted his album of the year award that he was convinced that his award-winning "Unplugged" album" wasn't worth releasing.
>
> "I didn't want this to come out," he said, "then finally agreed to it coming out in a limited edition. Then it sold a few, a few more, and I thought, 'Why not give it a try?'"

I don't know about you, but that beautiful, haunting, evocative song can still well me with emotion when I hear it more than a decade after it came out. Yet, at the time, this giant recording artist was unable to recognize the power and reach of the melody and lyric he composed and performed. Perhaps it was his grief and the personal nature of the song having been written after the tragic death of his young son. Whatever blocked his vision for the potential success of this work, it pointedly serves as a reminder as an artist you never know how things are going to work out.

Because you are unsure, or conversely, because you are positive your newly submitted pieces are going to set fire, don't cash any checks until they arrive and accept that sometimes things don't happen the way you would desire. And think of Eric Clapton when, in your mind, you are unsure of the value of the commercial appeal of your work — you might be surprised.

Over the years, artists often have asked me my opinion if their images were ready for market. Most often, I could or would not reply one way or the other. Naturally, I liked some more than others and there would be some that, regardless of whether they appealed to me personally, I could see had the "right" look for the current market trends.

I will tell you how I responded then, and would now, to the question of will your art sell. "The only people whose opinions matter are those who are willing to fork over some dough to buy your images." Just as with Clapton and his fans, although he couldn't tell when the album was recorded, they resoundingly voted with their pocketbooks and sent the message to him through sales and that led to the industry's top awards for achievement by a recording artist.

Everyone else, you, your publisher, the retailer, can all work hard to make the right impression and influence the buyer, and all that activity will help. But, in the end, the consumer has to want what you have created.

Building a dealer network one at a time is what creates the selling base. That base forms the platform for an art sales phenomenon to take off. When phenomenon takes hold it is driven by measures outside the control of the marketer. It takes a life of its own. In his bestseller, *The Tipping Point,* Malcolm Gladwell explains how seemingly insignificant incidences or a series of them can lead to phenomena very well. An interesting recommended read.

Marketing success and phenomena are the reason word-of-mouth is the most important aspect to the success or failure of motion pictures. How else can you explain why the quirky, relatively low-budget indie film "My Big Fat Greek Wedding" goes off the charts, and in the same summer, the big budget Steven Spielberg and Tom Cruise thriller, "Minority Report" rolls craps.

In the end, the buyer is in control and no amount of high profile appearances and marketing dollars will make a huge effect on people's opinions. If you are lucky enough to be painting the daisies or angels and they become all the rage, you are going to see some good sales result from the effort.

> The early bird may get the worm, but the second mouse gets the cheese.
>
> – Stephen Wright

The stories in the industry abound with artists who claim to have been the first to come up with a concept or theme that then takes hold in the popular market. The most well known marine artists like Robert Lyn Nelson, (*www.robertlynnelson.com*), Christian Riese Lassen, (*www.lassenart.com*) and Wyland may debate and lay claim over who was the first to portray the thematic over and under ocean view concept. Nevertheless, it is more likely true that California artist George Sumner (*www.sumner-studios.com*) as the one who actually pioneered that theme. While the debate over who fathered the idea can rage, the reality is the three former artists all have benefited financially from the concept far more than has George Sumner.

The previously mentioned quote from the *Time* magazine article about Thomas Kinkade thinking about throwing in the towel on painting cottages seems shocking given the man's eventual record-breaking success in the print marketplace. It's interesting then to note that California artist Marty Bell, *www.martybell.com*, lays claim to the concept of painting romantic cottages for a collector society years before Kinkade decided to try them. The dates on her first cottage prints substantiate her claim.

Still, Kinkade also decided to paint cottages and even to build a successful collector society around them just as Marty had done. The results of that decision made him the most financially successful print artist ever.

If you notice a trend for low-slung pants for young women is in, you can bet every pants manufacturer whose market is young women will have that look in their lineup as soon as possible. Why then should it be any different in the art business?

It has been said since Shakespeare there have been no breakthroughs in the basic concepts of writing plays, e.g., comedies, dramas and so forth. In one way or the other, all writers since have been creating permutations of his writings. I'm not a Shakespeare or English Lit scholar, but the point is all artists borrow in some way and have been influenced by artists that preceded them. We don't live in a vacuum.

It's understood your art comes from a deep personal creative well and you don't want to share from that well. Who would? What happens though is your art becomes part of a larger thing around which a business is organized. It is ultimately recognized not just for it's beauty and aesthetic quality by those who purchase it, but also by marketers in the form of competing art publishers as a fashion look and potential cash flow generator for them to emulate.

What does this mean? Being the creative force behind a new look or trend is not a guarantee you will be rewarded for your efforts. It depends on how fast your competitors start knocking you off, or if some of them are able to refine your look and make it even more popular than the original. It also means you can be on the other side of the equation and take cues from your competitors and your publishers to create and be influenced by what you see in the marketplace.

You can bet any publisher worth his salt is constantly on the look out for what is "hot" as not to miss some market share when the next "angels", "daisies", "urns & ferns", "martini-themes" or whatever takes off. You also can anticipate when they see something that looks as if it would fit into your style and capability they will suggest it to you to see what you can come up with for them.

If your vision is strong enough and your creative output prolific enough, you can own a look regardless of how many copycats come out of the woodwork seeking to take a share of your market. The great Americana artist, Warren Kimble, whose work has been a stalwart for years for poster publishing industry juggernaut Wild Apple Graphics, *www.wildapple.com,* has spawned countless imitators. Some undoubtedly have crossed the line only to realize how vigilantly Wild Apple defends its copyrights. Yet, none of them have come close to approximating Kimble's ongoing and storied success.

If you are going to work with a publisher, you really couldn't do much better than aspiring to define and own a look for years on end as Kimble has done. The trick is he is an original and being able to do what he has done with yet another look and style is difficult. Still, one can look at Kimble's work and see the influence of Jasper Johns and other artists whose works helped to shape and influence him.

You are faced with the same advice as for would-be self-publishers in that you need to take a brutal assessment of your talent and capabilities. By keeping that assessment in mind when you begin to envision how you would want to see your own success working with a publisher come to fruition, and if you have established for yourself realistic, attainable goals, you undoubtedly are going to enjoy the results of your efforts.

> Take the attitude of a student, never be too big to ask questions, never know too much to learn something new.
>
> – Og Mandino

Chapter Ten
Trends and Inspiration

> **There is no abstract art. You must always start with something.**
>
> **– Pablo Picasso**

To be frank, it seems presumptuous to have a chapter titled "Trends and Inspiration" in this book. That is because I don't think I can tell you where to find true inspiration. I think at the most elemental level it is part of the soul of being an artist.

What I mean by having inspiration comes in part from my own observations as a fine woodworker. I found many members within the St. Louis Woodworkers Guild, where I was president for two years, were as talented at their craft as I've ever seen. But regardless of their talents, an obvious division existed among them that was never openly discussed by the members. Nevertheless, it was discernible by those paying attention.

Many of our members, who were capable of building anything from wood, had no interest or confidence in designing their own projects. They would scour the woodworking journals and plan books for ideas on things to build. They would borrow plans or sketch and measure pieces, and recreate them, but you never saw an original design from this faction.

Others were always busy designing their own projects from scratch. Once the pieces were finished, the origin of the design was not an issue. The interest was in how the piece looked finished and what went into its construction. If you will recall the woodworker's comment in his letter to the editor, "The first one was fun to make." Obviously, he got more satisfaction from the creative challenge of making his furniture from new designs than he did from rebuilding and finishing more of the same pieces again.

I have noticed some similarities with visual artists and woodworkers and have had several publishers comment on this to me as well. They know some artists who are talented and can paint in a variety of styles and mediums. The catch is they possess the technical talent, but they lack the vision to know what to paint.

Most publishers are looking for artists who can bring them a look. Even then, poster publishers in particular often ask the artists if they can make changes to the images to give them appeal that is more commercial. Of course, that choice is always up to you; it is your copyright and property.

The bottom line for you as an artist is you need to develop a recognizable style so collectors, galleries and other buyers can grow an appreciation for your work. Some of you might already have found the style that is or will be your hallmark. Others of you are searching for something that you can hang your hat on.

Finding that "something" is what this chapter is about. Where do you look for trends and inspiration? The good news is you already know most of the sources. You already have learned that being an early adopter is sometimes not the optimal place to enter the market. It doesn't hurt to be an early adopter if you have just the right look and can harness the marketing to make a move with it. If you do adopt early, try to emulate the terrific slogan employed for years by the Panasonic Corporation, "Slightly Ahead of Our Time." It is perfect for the art business, too.

For those of you who are looking for trends, you can go right back to your trade magazines and tradeshows for great inspiration. You'll see what your competitors believe are the hottest looks in color, sizes, prices and content. That resource couldn't be more specific in terms of being helpful. Again, I encourage you to be a student of the business.

You need to go beyond the art print business to really become immersed in what is happening in the world of color. *DECOR* publishes a report every year from the Color Marketing Group, *www.colormarketing.org,* usually penned by its longtime columnist, Vivian Kistler. Vivian also is a member of the organization that bills itself as the Premier International Association for Color and Design Professionals. What they attempt to do is forecast the colors that will be popular in the coming year.

What you often see happen is the colors they predict actually start to gain traction with consumers about a year later. The furniture industry relies heavily on this group's forecasting to help them make choices about what colors will be going into their products.

If you were creating art for the poster and open edition print markets, as you plan your images, you would be wise to take your cues from this group. You see, by the time the furniture manufacturers respond to cues from the Color Marketing Group, other influences, gets the ideas into production and finally to your local showroom, it can easily be a full year later.

Have you ever wondered why lime green and burnt orange are coming back or sea foam green and peach are on the crest or why jewel tones seem to be the rage? It is often because color professionals have determined these shades are what will bring in the best return for manufacturers of furniture and home accessories. Wall decor, as it is known in the furniture business, is part of that mix. The more your imagery matches in color with what those buyers are looking at for the rest of their line, the better your chances a volume framer is going to put a large order in for your work.

If you are thinking that your art transcends the couch and you are above being influenced by mundane things, that's an artistic decision, but it's also a business decision. Even if you are creating art intended for higher priced giclées or serigraphs, it can still help to know what influences are driving consumer buying habits. You don't have to include burnt orange into your palette just because it is popular, but it won't hurt you to let the knowledge of it influence your work in some subtle ways either. On the other hand, if you are after the wall decor market, you want to dial in the right color combinations. As with all the advice proffered, it's your choice.

Unless you are averse to shopping, or even window shopping, you can make a working field trip out of visiting the big lifestyle retailers including Crate and Barrel, The Bombay Company, Pottery Barn, Pier 1 Imports, Macy's Home Stores and Z Galleries. Add in visits to your better local furniture stores to round out the excursions.

Other great sources are the catalogs from Horchow, Gumps, Frontgate and other top home-furnishing catalog merchants. All these influences will be a huge benefit to your ongoing trend education. Now you can stop throwing away all those catalogs without looking at them and take some time to observe what the color trends are in them, especially those for clothing, home furnishings and accessories. Other areas that affect color marketing are paint and apparel. Paint is more of a following trend, while fashion apparel is a leading trend.

A good way to see what is popular is to take some time to view the model home decor in your area. The designers who put together those sales house showcases are usually right on top of what is hot in the business. Often they are taking their cues from some of the same industry pubs that you are reading, plus others that are specific to their business.

Designers also are reading shelter books, as they are known, including *Elle Decor, Home and Garden, Metropolitan Home, House Beautiful, In Style, Architectural Digest* and more. You don't need to subscribe to all these books; you can review most of them in libraries for no charge. Many libraries will let you check out past issues.

These days even Target is into the act by hiring hot shot designers Michael Graves and Isaac Mizrahi. With the help of these designers, Target has put fashion and color trends into the mainstream, so you can even look for ideas the next time you're shopping for a 48-pack of Charmin.

You can find trend indications from the mainstream by watching some of the top rated sitcoms and dramas on television. "Miami Vice" with Don Johnson and Phillip Michael Thomas outfitted in pastel with their office and the show lighting also done in pastel, reflected the colors and helped define the look of the 1980s. The mega-hit "Friends" show with its prominently displayed Vintage Poster on the set every week helped fuel the rage for that look and format that peaked in the late '90s. The Cosby show helped Black Art go mainstream when the cast visited a gallery that carried art by Black Artists.

A surefire way to catch on to trends is to watch any of the cable shows on designing for the home, as those seen on the House & Garden channel (HGTV). These shows are loaded with ideas on topical design and color themes — the same things that are influencing buyers at the retail level and at the volume framer (OEM) level.

Keep your eyes open for any remodeling of local malls, or the construction of new malls and office buildings. Observe what is being done with them in terms of their style and color schemes. It can make some everyday things that would normally go unnoticed turn into something that can be fun for you to monitor.

Content or subject matter is always going to be easier to pick out of the pages of your favorite trade magazines than from the decor of your local mall, but there are still other resources to consider. One terrific source is Lieberman's Gallery, *www.liebermans.net*. It is the largest wholesale distributor of art prints and posters in the industry. It is known as a consolidator.

Its business model works this way. Typically, poster publishers in particular have minimum orders in terms of number of units or price before they will accept and ship an order. If Barney's Poster and Frame shop needs one or two copies of the latest Jack Vettriano poster that is distributed by Image Conscious in the U.S., it might either have to make a larger order with Image Conscious than it needs at present to make the minimum or forgo ordering until it can be met.

What Lieberman's does is allow small shops to order one or two prints, or even one or two prints from several different publishers. Then it places a consolidated bulk order with its publishers, receives them, and repackages them to be sent out to the retailers.

The price is the same to the retailer because Lieberman's s a volume discount the retailers can't attain.

Everybody wins in this scenario. The poster shops can keep selling one print at a time from their catalogs without having to stock things they don't have the money or confidence in to put into inventory. The poster buyer gets the order filled sooner and the publisher moves another unit out of their inventory, too.

This remarkable business model has become so successful that Lieberman's is a most important customer for many publishers. This also means it not only aggregates orders; it also is in the unique position to know what is selling best. Conveniently, the company occasionally updates its Website with information about what the bestselling images are so you can see what is popular. It's another source for you to work with when you are looking for trends and inspiration.

A new development for Lieberman's, announced in March 2005, is they are now distributing limited edition prints as well as posters. It will be interesting to see how this dynamic plays out in the art market. Can Lieberman's rise to become the number one or two customer of limited edition publishers as it has for many poster publishers? Will this help create more sales for everyone? Does this mean that some limited edition print publishers are looking to change their dealer gallery model of distribution?

I suspect is will be a good move all around. I base that statement on the reputation that Lieberman's has earned over the years of being a reliable partner for both the retail and publishing customers that they serve. They could not have grown to their preeminent position without having built a great deal of trust in them along the way.

This move also portends again that the business is changing and that those companies affected by the changes aren't sitting around waiting to become extinct.

They are actively reacting to changing conditions and looking to create opportunity from the chaos it causes.

When you begin working with publishers, they will be able to help you with ideas, as their resources are greater than yours. Often, their staffs attend the IHFC shows in High Point to look for trends and for new possible volume customers. When you put it all together and start thinking strategically about what you'd want to paint and what you also think will have an impact in the marketplace, you are getting closer to obtaining the success that is possible from this end of the business.

If you can find a way to regularly read some of the trade publications that serve the home furnishings field as *Accessory Merchandising* and *Home Furnishings News*, and so forth, you will have yet another valuable source of inspiration for trends and ideas.

> I have no data yet. It is a capital mistake to theorize before one has data. Insensibly one begins to twist facts to suit theories instead of theories to suit facts.
>
> – Sir Arthur Conan Doyle (Sherlock Holmes)

Chapter Eleven

Websites and Email Marketing for Artists

> Remember that the most difficult tasks are consummated not by a single explosive burst of energy or effort but by the constant daily application of the best you have within you.
>
> – Og Mandino

While the debate over whether Websites make sense for artists still rages in some quarters, new evidence shows the Internet is radically changing things.

Like it or not, the Internet is causing all manner of artists to reconsider how they get the product of their creation to market. The artist formerly known as Prince signaled a change. To some it seemed a defiant act of career suicide at the time, when he abandoned traditional distribution channels, i.e., recording labels and music retailers in favor of selling direct to his faithful fans through his Website. It turned out to be such a great move for him that he became Prince again.

The marketing tactics used by Prince and other artists who used the Internet to reach their fans are worth noting. The Grateful Dead are the most downloaded and traded musical act in history. They proved a multi-million dollar enterprise could be grown on the strength of giving away their product by freely allowing copying and distribution of live performances. There typically is a special section at Dead concerts where tapers can hook up their equipment.

In the process, the Dead created legions of dedicated fans, many of whom spend money on tickets, authorized recordings and other ancillary and licensed products to more than makeup the loss in income from the free distribution of recorded music. The actions by these artists have helped to create an ongoing tectonic change in the distribution of information and entertainment and literary and visual art.

The Internet is allowing visual artists, writers and musicians to create product and sell it directly to the public. This dynamic change is only beginning to be grasped by all parties involved, including consumers, retailers, distributors, and publishers.

These days, consumers are doing their best to avoid traditional distribution. They are listening to music on the Internet and over satellite broadcasts. They are buying music downloads at Starbucks and over the Internet to play on their iPods. To avoid annoying each other with phone calls and emails, they laboriously type in Short Text Messages on their cell phones that can be read at the convenience of the receiver. They are avoiding commercials and time-shifting television programs using digital video recorders. The number of ebooks published every year is exploding. Just a couple of years ago, the idea for me to write and publish this book using anything but a traditional distribution channel was out of the question.

Artists are frustrated and attempting to take control over how their products reach their collectors, readers and listeners. This is not a time to be content with the status quo. Here's a cogent quote summing up that notion:

> Even if you are on the right track, you can still be run over if you are not moving.
>
> – Will Rogers

The inestimable Will Rogers could not have said it better. Taking note of these changes is the job of the trade press and that is why in their March 2005 issue, *Art World News* carried a front-page story on the Internet and pricing. The editors of AWN brought together some of the industry's top art gallery owners and managers to do a roundtable discussion on how things are changing in the industry.

One noted that his gallery is carrying fewer prints, which is even more interesting since he is also a publisher of an artist in the print medium. His rationale for carrying more originals is they do not have the same competition as prints. The roundtable brought up the fact collectors are harder to sell to on the first visit than ever because the Internet makes it possible to price shop and to do due diligence on the gallery's claim about the artist's reputation, the art's provenance and so on.

Apparently, there also are price wars between galleries carrying the same prints by an artist. While this has always gone on, the Internet is exploding this shift in buying tactics. Smart collectors are realizing they can shop fine art prints as with everything else thanks to the convenience of the Internet. This is a problem for artists and galleries and at the same time, it is an opportunity. It's pure Darwinism. The survivors will be those that learn how to adapt to shifting consumer-buying trends. The same will be true for artists. It should be no surprise publishers are working harder at helping galleries find customers for their prints.

E-marketing to Consumers

I mentioned earlier it was my belief an artist should not create channel conflict with his or her galleries and I stick by it – unless you are a brand. Why would a gallery give up valuable retail space to compete with an unknown artist who is selling the same work on his or her Website?

Selling direct to collectors can be a viable marketing decision when you are a known entity. For example, the premier primitive artist, Jane Wooster Scott, appears to do well selling from her Website, *www.woosterscott.com*, while supporting a dealer network. You can aspire to be able to do this, but the recommendation is to have built your reputation and demand for your work first.

In Wooster Scott's case, galleries with Americana collectors will want to carry her work despite the channel conflict because they know she will sell in the gallery due to her reputation and demand for her work. It's a business decision for them as it is for her company. Still, galleries will steadfastly reject artists without a following that want to compete directly with them and who can blame them?

Whatever the case, I do think publishers, artists and galleries may quickly need to begin to rethink how they are collaborating in creating a collector base for the artist. What worked in the past may not work best now or in the future and the changes seen in shifting buying patterns caused by the Internet are at the heart of that rationale.

It's far too early in these nascent developments to predict how things will shake out, or even to flatly advise that one course of action is better than another regarding how artists, publishers and galleries will use the Internet to market art. My advice is stay tuned in and stay aware for opportunities and problems that can occur in sea changes as the ones we are seeing today.

There has always been some naturally occurring antagonism between some galleries and some artists. It's a marriage of sorts, and some are rocky while others solid. Artists feel galleries don't do enough for them and galleries think artists have no loyalty and are ready to jump to the next gallery, withholding better pieces and so it goes.

Of course, there also are wonderful stories of galleries and artists sharing blissful success in long-term relationships. But, I've heard artists sneer that in what other business can a retailer get their product for free and then keystone (100% price increase) it when it is sold. And, I've heard gallerists lament that it costs them $100 or more per month per piece on display in rent, not including other expenses as commissions, benefits, taxes, utilities, advertising, et cetera with no appreciation of their overhead from artists.

Galleries often rightly find many artists are clueless when it comes to understanding the difficulty of creating a collector. Artists often find galleries to fail them in properly representing them and living up to the spirit of their agreement. This is not a condemnation of either artists or galleries or their actions. Everybody has self-interest first which is natural.

I think this is not a good time for either side to be working against the other. The best thing that could happen would be for all parties to realize that if they don't find a way to cooperate better within the changes that the Internet is causing, that they will all suffer in the end.

My advice is if you are an artist, you need a Website. You will have to figure out how to use it to serve your own needs and that of the galleries or publishers who carry your work depending on circumstances. If you are gallery owner and are not setup to sell online and don't plan to be, then start thinking about what will be your next career as you watch the erosion of your business that goes to your online competitors.

Print On Demand & Affordable Giclée Priners

New technology developments will continue to cause change in the art print industry. Epson, *www.epson.com*, Hewlett-Packard, *www.hp.com*, Colorspan, *www.colorspan.com*, and the Roland DGA Corporation, *www.rolanddga.com*, are aggressively marketing affordable printers to the art market. Much like desktop-publishing software roiled the graphic arts market a few years back, the price of a digital fine art printer has come down to the point of being affordable for artists and publishers. Gone are the days of spending $100,000 and up to get your finicky Scitex Iris digital printer running properly.

These manufacturers are out to make the fine art digital printer ubiquitous as color inkjet printers. At least to the point if you need one, you can have it without having to take a second mortgage to finance it. No wonder some gallery owners are moving out of the print business – it seems like everybody's a printer these days.

I wouldn't be too worried about the development of everyone having printers. Just because someone can print, doesn't mean they have the smarts to get their work to market. That's where studying this book and following its suggestions can give you a decided advantage. There will always be a need for prints in the marketplace. What will be different is where and from whom consumers purchase their prints. That is changing too as big box retailers take market share from the small fry shops.

If you are thinking ahead that you could find yourself in the retail business either with a virtual gallery or brick & mortar, or both, you could well be right. Even better, it might be the best thing for you. Having has worked well for both Kinkade and Wyland to name two of a growing list of artists who market themselves all the way through the distribution cycle from concept, printing, and sales to collectors. Their work is popular enough that independent galleries want their images and are willing to compete with the company-owned galleries.

Publishers are still going to need product to push through their pipeline. They are always going to have access to markets that aren't open to individual artists, so while they are feeling the need to be light on their feet, they are never going to stop looking for the next "new" look that will jumpstart their sales. If you are lucky, it could be you. If you followed some of my advice on developing a look and keeping abreast of changing tastes in color and design, it could be you made your luck more than it happened to you.

A development running through the arts is called, POD (Print On Demand). I'll cover later about how Art.com is employing this in their strategy. POD means that artists, musicians, writers or even publishers can now create an image, write a book, record a CD and not have to inventory them in order to make them available to their customers.

To see an example of how POD works, check out *CafePress.com*. On this site, anyone can create a store where they can upload images and have them available as cups, t-shirts, sweatshirts, posters, cards, hats and more. Artists can upload music and have CDs created with silkscreened CDs and four-color jewel CD cases. Writers can upload PDF documents and they will print them in wire-bound or paperback covers.

What makes Café Press appealing is NO CHARGE to upload the art or to setup a basic store. The charge for a premium store is only $5 – $7 a month. Its pricing starts for making one item and goes down for quantity, but you pay nothing until an order is received to have the product produced. Plus, you set the price that you want to charge over the cost of production.

Keep in mind when you visit the CafePress.com, you are looking at the future. It might not be your reality yet, but it could as improvement in quality, speed, and range of offerings all become possible. There is likely nothing to complain about their quality now, but they don't produce artworks to rival any serious publisher now either. That's not what is important anyway. The model of what they are doing is important. It portends the future. The iterations of CafePress.com model that follow will be the ones to focus your attention. Lose that focus at your own risk.

In the mid-20th Century, merchandise from Japan was considered inferior junk and often it was; now look at what that country produces. In the latter part of the 20th Century, goods produced by Chinese manufacturers were looked upon as inferior. Now 6,000 of Wal-Mart's 8,000 suppliers are in China. How do these developments relate to you? Simply, if you are in the game for the long haul and not paying attention to technology and shifts in the distribution in how art is printed and sold, you'll miss opportunities, or worse find yourself on the outside looking in.

This POD technology is now starting to come into its own. Where it takes us as artists and consumers is anyone's guess. My guess is that those who take the time to figure out how to include technological advances as POD are going to have an easier go of it into the future, especially those who want to exercise as much control as possible over their careers. POD is behind the multi-size giclée editions you see offered these days.

Websites

I will admit my biases toward technology and the Internet come from having been the Director of Strategic Alliances for *DECOR* and my being involved with experimental projects as Decor Websites and Decor Expo Online.

Those experiences have made me more aware of the possibilities that Websites present to artists. As I see it, the case for an artist having a Website is strong.

The cost of a Website is lower than it ever has been. When you consider it might easily cost you $600 to design, produce and print a quantity of 500 unfolded 2-sided four-color brochures, a $300 annual hosting fee expense and a one-time $99 setup charge pale in comparison. Unlike a printed brochure, a Website is a dynamic thing, meaning you can update it as often as you wish. When a printed brochure goes on press, the opportunity to further update or change it ends.

That is not to say there aren't excellent reasons to use printed materials, because there are. Your Website should be your primary source of providing contemporary and continuous information for your collectors and prospective collectors, and for your galleries and prospective galleries. Websites offer an excellent and cost-effective way to consistently keep collectors, buyers and others up to date with what it is you have to offer.

The number of companies that offer Web hosting is huge. Far too large to list or try to evaluate in this book. I've mentioned *www.mysmartmarketing.com* as one resource that specializes in Websites and email marketing for artists. The company is owned and operated by Dave Beckers.

He has been in the art industry for more than 20 years. Through his company, Art Information, he offered the first digital compilation of prints and posters to galleries. Before the advent of the PC and the CD, Dave had images burned on to the platter sized Laser Disc format that he sold with a television. In the age of DVDs and broadband Internet connections, it sounds funny now, but in the 1980s, it was incredible that tens of thousands of images could be cataloged and presented through a television screen.

Dave sold Art Information and founded My Smart Marketing. He will build you a Website for $99 and host it for $24.95 a month, including your domain registration. There is hardly a reason not to do this at those rates. His company has a large email list of galleries and dealers and it can also help you with your marketing in addition to your Web building and hosting needs.

Another art specific Web hosting and building provider is a Canadian company called Art Affairs, *www.artaffairs.com.* At more than $50 per month and up, their sites are fancier and more robust but pricier than those of My Smart Marketing and they require a 2-year commitment. A new twist at Art Affairs is that the company founder, Meir Gluzberg, announced in March 2005 that he is buying back the company and will be running it. With his passion for the business, it will be interesting to watch for how his return influences it.

You can find many other low cost providers, even free sites from your ISP and so forth. But you get what you pay for. When something is free, you usually have to put up with ads and banners on your site that are not related to your site, or from which you derive no income. From my point of view, that's a cheesy way to display your art as a professional.

If you can find someone who is Net savvy to help you make decisions about where to build and host your site, by all means take advantage of that relationship. Otherwise, you'll find it is worth the time to learn enough so you ask good questions about your service provider.

Here are some important questions to ask a potential Web host:

- If they have an annual contract, can you get out of it with a money-back guarantee for time and services not used?
- Do they offer an uptime guarantee so that in the case of a serious outage, are you compensated?
- What backup system do they use, i.e., tape, CD-Rom, or other method?
- Do they have redundancy in their backup scheme?
- Do they have redundant backbone connections to the Internet, meaning they are connected to the Internet by more than one provider or server and are they major backbone providers?
- Do they offer 24-7 technical support by phone, or is it only via email?
- How long have they been in business and who are some companies they currently host who will offer references?
- Can you pay month-to-month or do they require an annual contract?
- Is it easy for you to upgrade/downgrade services you may or may not require as time passes?

- Do they host adult or other controversial content that would make them the target of attacks or prosecution that could cut service or access to your site in the process?

These questions will get you started. You will have other questions that relate to your individual situation after you not do more research on the question of where to host your site.

When it comes to questions, here are some concerns buyers who visit your Website will have. These are consumer concerns that you should be considering as you plan to launch your own Website:

- Does the seller's Website post a Privacy Policy?
- Does the seller have a return policy?
- Is the company name/owner shown with complete address?
- Who takes the risk of transportation damage or loss? Is transportation insurance included?
- Are all costs, especially shipping, spelled out?
- Does the site offer secure transactions? Meaning do you protect sensitive data by encrypting it when sent over the Internet using standard SSL (secure socket layer) or other appropriate technology?

There is no better or more cost efficient vehicle than a Website to begin to get exposure for your images. You already have the bandwidth available up to certain limitations that you probably can't exceed unless you are putting up hundreds of images, so there is no reason not to use it. Think of your Website as your updatable electronic brochure.

I believe that even if you find yourself in the comfortable position of having a publisher who is contracting for all your images, you should still have a Website, preferably one in your name. Remember, you are the brand. If you are Bill Smith, you are going to have to be creative with something as: www.billsmithartist.com.

There are many good reasons for you to have a Website. Even in the instance where all your work is going into print, your Website can be used to promote your personal brand, YOU. Use the site to sell the originals not marketed in other channels, or to take commissions and always to highlight your print images with links to your publisher or the galleries that sell for your publisher.

You have to use common sense and work out something that is agreeable with your publisher. Artists that have Websites that sell direct are creating what is called channel conflict. In Internet parlance, it means the publisher is competing with someone else that is distributing the exact same product. This frequently leads to bad feelings from the party who is feeling aggrieved and I can certainly understand why the publisher would feel that way.

Understandably, artists feel they need to protect their own interest and make as much money as possible. That's why it's important to rely on and work with reputable galleries to do the marketing and sales for you, or to get into the print market. If you feel you can't trust a gallery or your publisher to do the job for you, then listen to your instincts. They are probably signaling good reasons to move on.

As previously mentioned, the amount of art being sold on the Internet is steadily increasing. Examples are leading Internet art sales sites including: *www.art.com*, *www.barewalls.com* and *www.allposters.com*, whose frequent and significant orders to poster publishers continues to climb and this makes them increasingly more important as wholesale buyers and a growing source of revenue for them.

Next Monet, *www.nextmonet.com*, continues to find new print artists to present to their growing list of online print art collectors. Around since 1998, which in Internet or dog years is a lifetime, they have evolved into one of the top sites for online fine art prints and originals in the industry.

Something interesting to investigate is a program offered by, *www.art.com*. The company has begun an Original Art Program that coupled with the significant traffic the site generates, looks enticing. It offers several levels for artists participating in the program and the levels start with everybody's favorite price point — free!

According to the information on the Website, artists keep 100% of the proceeds from the sales of their originals. They have a POD (Print On Demand) program for those artists who are on a paying level, with the lowest starting at $50 annually. With that program, an artist is allowed to post up to 96 images in their "store."

If you are interested, you should contact the site and request more details about its printing partner and the POD program. As an artist, I think you would need more information before making a decision to participate or not. The point being, this is another example of how you could use the Internet to help advance your career.

You might recall me mentioning that Calvin Goodman, author of the *Art Marketing Handbook,* changed his tune regarding posters when he saw how well they worked for his client, Arthur Secunda. I believe many current naysayers, who dismiss the Internet as being an unfavorable environment for fine art and fine art prints will eventually change their tune, just as Goodman did towards posters.

It's not apples to apples, but nonetheless consider that Costco has become one of the largest online sellers of large diamonds in the world on their *www.costco.com* Website. Presently, stones priced up to $20,000 and higher are listed for sale. When I entered the term *"fine art"* into their site search it returned prints priced in the thousands by Chagall, Modigliani and LeRoi Neiman.

I'm not suggesting Costco is going to start carrying unknown artists anytime soon, but the fact it is selling expensive prints by some famous artists on its Website is another indication of how fine art print sales are being integrated into the Internet. If consumers will buy expensive diamonds online, it only makes sense that they would buy fine art online too, especially from artists who are a brand.

Website Marketing

If you have your own Website and are interested in using the Internet to help promote your art, you should spend the time to do some exhaustive research. Use Google and Yahoo to see what comes up when you put in search terms you think people might use to find your art. You can find a wealth of valuable information by going through this exercise. Since the same terms yield different results, you need to use both to get the best feel for what you want to learn and how it will apply to you.

By searching thoroughly, you will come up with lots of great ideas; free information and sites that will help you market. You will find competitors who are doing what you should be doing and a smattering of those whose primary intention is misguided, overblown or patently fraudulent. Choose your partners wisely and do your research carefully before you enter into any agreement with an unknown company on the Web.

You can read many books and more pages on Websites than you have time for on the subject of getting the right Meta tags and keywords worked into your Website as part of your Search Engine Optimization program. If those words all sound foreign to you, you probably need to hire someone to build your site and to help you market it. Sometimes it can be the same person. In other cases, you will need a different person to help you get your site submitted to Yahoo, Google, MSN and other Search Engine sites.

There is no denying it's a lot of work to do these things and you have to balance that with all the other things you need to accomplish, as in getting some painting done. It could be you will have to grow into your Website. You can start with something easy and uncomplicated and use it as your electronic brochure. As you become more comfortable, you can investigate the more advanced ways of marketing your art through your Website.

On the other hand, if you were born with a beanie hat with a propeller on top, you are probably already advanced beyond the skipping of the high notes this chapter covers on Internet marketing. For those of you who are somewhere between neophyte and certified geek, I hope to pass along some useful suggestions, ideas, books and sites to visit to improve your knowledge and results.

If you are using a Windows operating system, you can download a free program called "Good Key Words" from: *www.goodkeywords.com.* It will help you find out how popular any search term you might think of is on various

Search Engines. While this won't replace doing the above-suggested research, it will help you in the next step of marketing your art.

You can buy keywords on some Search Engines. The primary way is through a company called Overture, found not surprisingly at: *www.overture.com.* It works with Yahoo, AOL, MSN and several other large Search Engines and through it; you can buy keywords to help drive traffic to your site. You choose the keywords and amount you are willing to pay when someone clicks on your link through it to get to your site.

Yahoo owns Overture and it is likely it will be rebranded with the Yahoo name by the time you are reading this. Regardless, the Overture links will forward you to the newly branded Website when it changes.

Another program is Google AdWords. Since it doesn't work with Overture, you have to deal with it direct, but the process is much the same. The costs are about the same, although there are some differences in how the programs are administered. A recent news item indicated that MSN is considering its own ad placement service that presumably would replace what Overture does for it now. Stay alert, the landscape is changing as these Internet giants realize the importance of owning the companies that sell pay-for-click advertising on them.

On Overture, the bidding starts at as little as $.10 per click. At those rates 20 visitors being directed to you would only cost you $2.00. You do control how much you will spend and how much you will bank with them to keep your listings current, so there is little downside to trying them once you have your Website functioning.

Since Overture will let you go on its site and play around with all the keywords you can think of and find out what companies are paying for them without having to sign up for an account, I encourage you to give it a try.

When I recently checked the broad use keyword phrase, “fine art prints,” it was bid at $.99 for the top listing. These results mean you will need to be more specific with your search terms if you are on a budget. Add descriptors such as: giclée, floral, Southwest and so forth. Get even more specific if you can, but don’t get so specific there are no or few clicks on the terms you use.

Yahoo is now promoting its local listings. With that program, you can target only responses from people in your local marketing area. This can be a nice feature if local marketing fits into your overall plans. There is information on the Yahoo site about this service and many other features. It is worth spending some time there reviewing the site at: *http://smallbusiness.yahoo.com/marketing*

Targeting local audiences on the Internet is growing in popularity as the Yahoo initiative mentioned above testifies to. Here are some things you can do to refine your search engine optimization to target your local audience. This is important because as the Internet grows and Search Engines become more complex users have to sift through information and drill deeper on pages and links to find geographically desirable results.

- Make certain you include keywords and phrases that are geographically oriented. Do the same for title, description and your Meta tags

- Have your contact information that includes your company address and ZIP code on your home page

- List your company with all the online vendors of Internet Yellow Page kind of listings as:

DexOnline, SuperPages, and especially those that are local to your area. Search Engines run programs known as Webcrawlers or spiders to catalog pages on directory Websites far more frequently than they will yours. Having local directory listings will help improve visibility for your own listings.

- Look for local link exchange possibilities as such as coffee houses, restaurants, and other simpatico companies
- Check out *www.localsearchguide.org.* It is a wonderful resource for information and advice on how to use search engines to do local marketing and advertising
- Don't forget that the keywords and phrases your local audience uses to find you are key to your site optimization

Of course, if you are going to sell on the Internet, you need a form of payment. The easiest is Paypal. It's a safe bet many readers of this book have already bought something from eBay using Paypal. Setting up a system to be paid on your site through Paypal is easy, though not as cheap as having your own online Merchant Account if you have the volume to make one work for you.

A Merchant Account is an online credit card processing service. It can be, but often isn't linked to your brick & mortar Merchant Account. They have their own rules and charges that are completely separate from the online version. Sometimes you can get a volume discount for having both a landed and online account with the same provider. You really must shop and compare to get the right deal with a qualified company.

Merchant Accounts are great, but they are expensive to start and keep up if you don't have a lot of traffic. A Merchant Account is probably something you will grow into as your business online increases.

Another alternative to Paypal is 2Checkout *www.2checkout.com.* I use them to process orders for ebooks and services on my Website: *www.printmarketprofits.com.*

2Checkout is different from having a Merchant Account. They charge a one-time setup fee and 5.5% of the transaction plus $.45 per transaction. That might sound expensive and it is compared to half that rate for an online merchant account. However, unless you are doing volume business, merchant accounts are expensive and can cost you more in total fees than 2Checkout or Paypal, neither of which charge monthly fees.

Another way to get your art on the Internet is through what is known as affiliate marketing. This is where you agree to allow your items to be sold by others, or you provide links to your site to others and pay them a commission when a sale is made that comes from a link on someone else's site to yours. For instance, 2Checkout has an affiliate program with dozens of artists and galleries who sell through them.

There are companies that specialize in affiliate marketing. A prime example would be Commission Junction, *www.cj.com* where it has an extensive program of letting buyers and sellers find each other to promote the sales of their products and services. Other well-known affiliate marketing companies include, Link Share *www.linkshare.com* and ClickXchange *www.clickxchange.com.* If you are so inclined, you can find ways to work with them to make print sales.

A popular method of gaining traffic to your site is to make link exchanges with other Websites that have traffic that would be compatible with your site. This is different from affiliate marketing where you pay commission to sites that direct paying traffic to you. In a link exchange, you agree to place a link or banner on your site in return for getting similar treatment on another site.

Try putting the term "art link exchange" into Google or Yahoo and you will find many, many opportunities to create exchanges. Not all exchanges available are going to be worth your time or Web space to deal with. You will need to do some investigation before going forward with any companies you are not familiar with. I view this kind of activity as a secondary effort to bolster your Search Engine Optimization and email blasts to your qualified opt-in list, which should be your primary sources of generating traffic and sales on your site.

Other sites specialize in helping self-representing artists get their work onto the Internet. A Website that looks worth investigating to me is: *yessy.com.* Yessy charges a flat $59 setup fee and 10% fee per each piece that is sold. They have unlimited space for your art and accept credit card charges which are included in the 10% fee.

Yessy appears to have an aggressive marketing program with links popping up in Google Ad Words (paid ads on the right side) for appropriate search terms. Its Website shows more than 25,000 pieces on their Home page next to the Paintings link. That volume indicates support from artists. The photography category shows more than 20,000 pieces available.

You should also review *www.absolutearts.com* and *www.fine-art.com* for more examples of Websites where you can promote your art. There are so many more that it is not possible to find and report on all of them and I'm sure I've left some worthy out of this report. In my opinion, none of this kind of activity should replace having your own Website where you can control content and every other aspect of the site. These additional places are just to help you get more reach and that is how they should be treated, not as a replacement for your own Website.

Because these activities can be classified as advanced Web marketing and because they require experience and time, I advise you to wait until you have mastered the basics of getting your Website functional for its primary purposes before getting involved with affiliate marketing and link exchange marketing. It's still good to know about them now.

EBay

EBay is the virtual 800-pound gorilla, the Wal-Mart of cyberspace. It is a phenomenon with no comparison. The metrics, as geeks call the numbers, it puts up are staggering no matter how you want to look at it.

While every other dotcom that burst on the scene in the last ten years is either still running deeply in the red (Amazon) or valued astronomically beyond their intrinsic worth, (Can Google really be worth 5 times more than General Motors?), eBay has been profitable since day one. That's not to say the stock price is fairly valued; my guess is that long term, it's not, but I didn't write this book to give stock advice. I'll only leave you to your own devices with this admonishment from the renowned economist, John Maynard Keynes, "The stock market can remain irrational longer than you can remain solvent."

There is a book you might want to add to your library, *Internet 101 for the Fine Artist with a special guide to Selling Art on eBay* by Constance Smith and Susan F.

Greaves. I haven't read "the guide" yet, but it looks interesting from its description on Amazon.

Smith's Website, *www.artmarketing.com* hosts the site *www.workingartist.com.* It is on that Website where you can use the previously noted free software to keep track of your contacts and the history of your artwork. It also offers a wide range of mailing lists that are of interest to artists on the site.

I recommend you also check out the sites *www.artsiteguide.com* and *www.ebsqart.com.* Both offer help for marketing on eBay for the advantage of self-representing artists. The former provides a host of links to other sites helpful for marketing art on the Internet. The latter is one of the older sites on the Internet designed with the purpose of helping self-representing artists. Both have ties in one way or the other to eBay. Explore them for inspiration on how you can use eBay in your marketing mix.

The sheer number of Websites devoted to fine art and artists is staggering. To attempt to list them in this book would be counterproductive. Nevertheless, spending time with relevant searches on Google and Yahoo (yes, you will see a different order of results) is an excellent way to get educated on what is out there. I strongly urge you to do so to broaden your knowledge about the growing online competitive market in art print sales.

I read a magazine article in Home-Based Business magazine on artist, Natasha Wescoat, *www.natashawescoat.com.* Good for her, she is using her promotional muscle to find new markets. She sells through her eBay store and has become a PowerSeller for them. According the following language on eBay:

PowerSellers are eBay top sellers who have sustained a consistent high volume of monthly sales and a high level of total feedback with 98% positive or better. As such, these sellers rank among the most successful sellers in terms of product sales and customer satisfaction on eBay.

When I viewed, she wasn't showing prints on her site, but that is irrelevant as an artist could use the information she has put together to sell whatever medium they are creating in and still be as successful as she is. Since I see the name Michael Wescoat listed as the PR person on her site, I imagine she is one of the fortunate artists who have been able to enlist a spouse or family member to help her market her art.

Another artist selling originals online through eBay is Selsillie Girelli, *http://stores.ebay.com/Girelli-Gallery*, who says she is getting more than 1,000 hits per day on her online store. She compares that to 10-20 people who might visit the typical gallery in a single day.

I don't know the actual income being generated by these and other artists who have decided that eBay is their ticket to success. For them, it may be. For you, it is something you have to grapple with. The decision to use eBay must be taken into consideration in your career just as the decision, discussed earlier, by Yuroz to stay at ArtExpo has affected his career. And, as previously noted, there are no wrong decisions when it comes to these things. Only the right decision for you is what matters.

One of the first things mentioned in this book was the need to be as clear as possible with regard to knowing what it is you want from your career as an artist. One has to wonder, would an artist like Van Gogh who struggled to sell his work in his lifetime try to use eBay if it were available to him? Are there artists today who are being ignored by museums and critics and those in power positions of the fine art world who are selling online, and who someday after they are gone, will be discovered as a revolutionary genius? Far out concept? Undoubtedly, but life is always stranger than fiction, so you never know.

Email Marketing

One of the main purposes of having a Website, besides displaying your art on your own dynamic electronic four-color brochure, is to help you market your art to dealer through your Website using the tremendous efficiencies of email. Email is the most cost-effective method of getting your message out to a targeted list of buyers and collectors. While the volume of spam has reduced its effectiveness, it still has an central place in your marketing mix.

We talked earlier about the importance of building a dealer network and it is a process that needs to be worked on every day. The same applies to email marketing. One of your most vital tasks is to be constantly focused on harvesting email addresses from those interested in your work.

Despite all the justified clamor about spam, you will be pleased at how many people will still trust you with their email address when you promise not to use it for any other purpose but to contact them about your work. To accomplish that end, you should have a privacy statement with a link to it on your Website. Further, you should clearly state that you promise never to share or sell the email addresses you collect.

If you want some help creating a privacy statement, the Direct Marketing Association conveniently provides a wizard that is free of charge to use. Go to: *www.the-dma.org/privacy/creating.shtml* to find the Privacy

Policy Generator. This link alone will pay you for the price of this book. You could easily pay hundreds of dollars for a qualified person to help you build a privacy policy that is appropriate for your business. Www.truste.com is one of the main providers of privacy policies for Web marketers. Their logo on a site is adds credibility about that site. Their minimum charge for creating a privacy policy is $599.

Having a link to a privacy statement on your Website will make you standout as professional and courteous. It is also something that you can refer to when you are attempting to collect email addresses. If a person knows you have a posted privacy policy, it is just one more reason to trust you with his or her email address.

Most sophisticated Web hosting companies incorporate ways for you to collect addresses and to email to them as needed. For those who start with a budget Web hosting plan, that feature probably will not be available. But, there are companies that can help you.

A company previously mentioned, *My Smart Marketing,* has extensive email lists and it will build email templates for you and mail to its lists of galleries and publishers for an affordable fee. Using this resource is perhaps the quickest, easiest and simplest way to get started with email marketing. The lists are built on what is known as opt-in lists, meaning the addressees have agreed to have email sent to them. You don't need to worry about your incoming message being considered spam.

Another way to do email marketing is to use an email marketing company such as Constant Contact *www.constantcontact.com.* It has easy-to-use templates to help you build professional looking HTML emails. It separately sends text mails to addresses on AOL that usually have trouble receiving HTML email.

You can manage your emails by using the feature that shows you the bad addresses and easily remove them from your list. It also provides HTML coding which will build a box to place on your site to ask your visitors to sign up for email messages from you. When visitors enter their email and click, "Submit," they are taken to a page where they can check boxes that you design to indicate how they would like to be contacted. For instance, you could have a newsletter box, a new art release box, a press release box and so on for them to check according to their interests. You choose how elaborate you want to make the choices. My advice is to keep it simple with no more than three choices.

You may try Constant Contact free for your first 60 days. After that, for a small monthly fee based on the number of email addresses you have on your list, Constant Contact will allow you to send out unlimited emails to your list. It's free if your list is fewer than 50 names.

When you visit the Constant Contact Website, you also will be offered the chance to download six different White Papers & Guides to help you become a smarter and better email user. I have downloaded all of them and found many useful ideas in reading them. I'm sure you will, too. Here are just a couple of titles from their site:

- 10 Tips on Getting and Keeping Permission
- 25 Email Marketing Terms You Should Know

As with Website marketing, you can spend all your time trying to get proficient at email marketing. It's better to hire someone to help you or you can go at it slowly and try to grow your knowledge and capability to make better use of these wonderful tools. What I like about both Website and email marketing is that even though the learning curve can be steep, the payoffs can be worth the effort and you can grow at your own pace.

I've been around enough artists to know what their time is like and I don't have any delusions that hordes of artists are heading out to get on the 'Net while firing off robust email campaigns as a result of reading this book. What I do hope is that I've imparted enthusiasm and encouragement to each of you to include Websites and email marketing in your own marketing mix because I believe they will only continue to grow in importance and pay quick dividends for those who learn how to use them.

> A wise man will make more opportunities than he finds.
>
> — Irish painter, Francis Bacon

Chapter Twelve
Licensing

> **Business art is the step that comes after Art. I started as a commercial artist, and I want to finish as a business artist.**
>
> – Andy Warhol

The subtitle of this book is "Creating Cash Flow from Original Art." The concept is if you can create a multiplication process that allows you to make money more than once from your original art, it will create a nice secondary stream of passive income for you.

Imagine if the same images could be used in yet another multiplication strategy that created a tertiary or third passive income stream for you. How great is that? That is what licensing represents to you and your publisher if you are with one.

Licensing is the business of offering royalties or commissions for the use of art, logos, likenesses and entertainment characters, et cetera. Licensing can come into play when a graphic element is to be printed or placed on a variety of products as posters, note cards, wallpaper, linens, puzzles, playing cards, stationery, and much more.

There are a couple of books on art licensing already available and I will pass them along to you in this chapter. The point of this chapter is to introduce you to the concept, make you aware of how licensing works in general and encourage you to learn more about it.

Until the last decades of the 20th Century, most of the licensing of fine art was limited to publishers licensing original art from artist in order to make prints. You might not have even thought of having a publisher make your work into a poster as licensing, but it is.

To this day, having publishers create prints and posters of their art represents the majority of licensing deals for artists. And, the royalties for licensing fine art images to become prints remains the largest in the industry.

What happened in the 1990s was an explosion of licensing for all manner of things, including fine art becoming a consumable commodity of sorts when pulled into the mix. Do you recall the "Dancing Raisins"? If you do, that was about the time licensing took off. Once it started, you began to see the cross-promotional marketing between retailers, fast food companies and others with movie, television and comic book characters being seen on lunch boxes, pajamas, interior decor, linens and available as action figures in Happy Meals, and so forth.

In addition to books on the subject of licensing, there also are magazines and tradeshows devoted to them as well. License! is the premier publication in the field. It is published by Advanstar Communications, which also produces the annual Licensing show in New York.

There also is a British publication called, *Total Art Licensing,* but they do not come up with a Website in Google searches drilling down through more than 10 pages using the term, "Total Art Licensing." I presume the company doesn't have a Website.

Advanstar also produces one of the industry's most important shows, Licensing International 2005. *www.licensingshow.com* For artists and would be publishers, attending this show is valuable. Even though this is not as important to a novice in the art publishing business, it is something you should put on your agenda as soon as possible.

For many art publishers, the licensing show called Surtex, *www.surtex.com*, has become more important than Licensing International. When you read the information below taken directly from the Website, you will understand why:

Markets Represented:

Art and Design for Bath Fashions, Bed Linens, Cookware, Decorative Fabrics, Floor Coverings, Home Accessories, Home Apparel, Housewares, Infant/Juvenile Merchandise, Kitchen Textiles, Specialty Appliances, Tabletop, Upholstery Fabrics, Wallcoverings, Window Treatments

Profile of Attendees:

Manufacturers of Bath Fashions, Bed Linens, Decorative Accessories, Floor Coverings, Home Furnishings/Textiles, Home Apparel, Home Furnishings/Textiles, Home Storage, Juvenile Merchandise, Kitchen Textiles, Packaging, Personal Care, Private Label, Tabletop, Upholstery Fabrics, Wallcoverings, Window Treatments. Retailers: Bed, Bath & Linen, Department Store, Design/Lifestyle, E-commerce, General Gift, Gourmet, Home Furnishings/Decorative Accessories, Home/Garden, Mail Order/Catalog, Specialty Stores. PLUS: Buying Offices, Interior Designers, Licensing Executives and Wholesalers/Distributors.

Perhaps the strength of the Surtex show for publishers is that it is more concentrated on art and design than the Licensing International show, which covers the entire gamut of opportunities. Undoubtedly, the most important reason is that publishers are coming away from exhibiting at the Surtex show with more contracts for their artists.

If you are an artist working with a publisher, most likely that publisher already has representation at shows as Surtex and Licensing International 2005. If it doesn't, I would want to know why and what other licensing plans and marketing efforts it has for the licensing business. Most publishers worth their salt long ago realized the great opportunity that licensing presented to them and their artists and have full-time staffs concentrating on that aspect of the business for them.

A couple of books I can tell you about on licensing include: *Licensing Art & Design* by Caryn Leland and *Art Licensing 101* by Michael Woodward. Having the information they present in your library will be invaluable, particularly if you plan to self-publish. As with other publications listed in this book, if you find purchasing them is hurting the budget, consider borrowing them from the local library.

A review from one reader on Licensing Art & Design said that is what she did (borrowed the book from the library) after seeing tepid reviews for the book on Amazon. She goes on to say it was well worth reading and the one suggestion on protecting her design copyrights made it worth the price. Often that is the case with tightly focused books, readers need come away with one or two ideas that make the job of managing their careers work better. Obviously, if those ideas or suggestions are powerful, they offer great payback for the cost to acquire the books and in the time spent to read them.

Licensing is an area where having a rep will make all the difference. I said earlier that reps for selling fine art prints are at the bottom of the heap on ways to move the sales dial for self-published artists and gave reasons why. In the case of licensing, it is the opposite. Having a rep will make all the difference between getting lucrative contracts with companies that want your images and being on the sidelines.

Largely, the licensing business is all about the Rolodex — who do you know and how well. Those agents who have the right contacts will quickly get your images in front of the best prospects because they have the established relationships and a reputation that they are reliable, have good taste and instincts.

Because there is so much work to do to getting a publishing business off the ground, I recommend that all but the most determined and organized self-published artists seek an agent or rep to handle this aspect of the business, at least in the first few years. The usual licensing agent commission runs around 35%, but nothing is set in stone. It can be lower or higher depending on circumstances. Nevertheless, those agent commissions can be cheap if they bring you multiple deals.

Whether working with a publisher, an agent or dealing directly with licensing contacts, when it comes to deals you'll find you have little time to make up your mind. Often they are looking for a specific image and usually have more than one source for it. Those that dither find the window of opportunity closes on them while they waited to decide.

An aspect of the licensing business that you need to be aware of is, unlike with a poster or print where the artist maintains some artistic control in the printing process, you give up all control when your image goes to licensing. That means your images can be reversed, truncated, colored differently, made in different sizes and more. If it troubles you to think your art is going to be treated this way, you need to think about whether licensing is your cup of tea or not.

A company that specializes in art licensing is Porterfield's Fine Art Licensing. You can learn much from visiting its

Website. The site offers many articles you can read about the licensing business and you can look over the imagery of the artists represented. If you are so inclined, you can submit to that company for representation.

While most artists aren't getting rich on a few licensing deals, the good ones with good representation find that over time their imagery creates a nice steady stream of cash flow. What happens is the same companies come back to license the images for new products. Each generates its own royalty stream and when these deals begin to overlap and that is when the process becomes lucrative for artists. It gets better when multiple licensing deals with multiple companies kick in.

I mentioned Mary Engelbreit previously. It is said she has 6,000 license agreements for her images, all which help propel her company to its annual $100 million dollars in sales. While she and Thomas Kinkade have to be near the top of the heap in terms of licensing deals, just imagine getting a fraction of that pie. For most artists the idea of making $50,000 or $100,000 for licensing royalties on their images is just a dream. It doesn't have to be if you work hard and smart at including licensing in your creative and business prospects.

> One must still have chaos in oneself to be able to give birth to a dancing star.
>
> — Friedrich Nietzsche

Last Words

To finish reading a book as this one shows your commitment. And, although it's not a guarantee of success, it's a sure sign you are serious about achieving it, which is no small part of the equation.

If you are ready now to commit to a print career, my wish for you is to find inspiration, as do I, in the following quote from the great German poet, writer and scientist, Johann Wolfgang Goethe. Best known for his epic work, "Faust," he also influenced abstract painters, including Kandinsky and Mondrian with his non-Newtonian, unorthodox theory of the character of light and color.

I come back to Goethe's eloquent timeless words over again for the sagacious motivating encouragement they offer:

> Until one is committed, there is hesitancy, the chance to draw back — Concerning all acts of initiative (and creation), there is one elementary truth that ignorance of which kills countless ideas and splendid plans: that the moment one definitely commits oneself, then Providence moves too. All sorts of things occur to help one that would never otherwise have occurred. A whole stream of events issues from the decision, raising in one's favor all manner of unforeseen incidents and meetings and material assistance, which no man could have dreamed would have come his way. Whatever you can do, or dream you can do, begin it. Boldness has genius, power, and magic in it. Begin it now.
>
> – *Goethe*

About the Author

Consulting & Speaking Services

In 1988, Barney Davey joined DECOR and Decor Expo, the business magazine and industry tradeshow producer for art and picture framing retailers, as a sales and marketing executive. For more than 15 years, he made major contributions to the meteoric sales growth of the magazine and the shows.

As the rise in importance of the Internet evolved, in 2000, Barney added the position of Director of Strategic Alliances. He used that platform to help foster growth with dotcoms and other companies seeking to form mutually rewarding relationships with DECOR and Decor Expo. He launched Decor Websites to offer easy-to-use affordable template-based Websites for retailers. He was responsible for starting Decor Expo Online, the industry's first virtual tradeshow.

Along the way, Barney enjoyed a bird's eye view of the print market as he consulted with hundreds of self-published artists and established publishing companies on the most efficient and effective ways to market their art prints to retailers. Barney has developed many friendships and close business relationships with the industry's leading print publishers. His observations with regard to their common traits and best practices are the basis for this book.

Barney is the son an accomplished fine artist and a serious fine woodworker in his own right. He is the former president of the St. Louis Woodworker's Guild. He has spoken to numerous artist groups around the country, leads art print marketing tele-classes for Artist Career Training as an art print marketing expert. He resides in Arizona where he practices art-marketing consulting, writes and does public speaking.

Consulting Services

Barney Davey provides clients with consulting on art print marketing, e-publishing and media sales. Work with clients primarily is done over the telephone and Internet. No travel means more time for you with out having to leave your studio, office or home for your session. All consulting is done directly with Barney Davey.

Artist Consulting -Typically, artists find that either a one-hour session or a three-hour session is sufficient to either help make crucial first decisions about self-publishing versus using a publisher, choice of media and assessing resources or to devise a workable strategy to further progress their art print career

One-hour Session - this session is suggested for artists who are beginning to investigate the art print market but have not decided how to best enter into it. That is, they haven't concluded whether to self-publish their work or seek to work with an established publisher.

Usually, the one-hour session is sufficient to help artists determine if they have the necessary resources to self-publish. Having a clear understanding of what it takes to self-publish and doing a personal assessment of their available resources can deter costly mistakes at this juncture.

Three-hour Session - this session is suggested for artists who have already made an appropriate determination regarding whether to self-publish or seek a publisher. Often, this artist has already been working at least part-time, if not full-time on their art career and is looking to move to higher levels of recognition and prosperity.

To facilitate an optimal outcome a goal for both the one and three-hour sessions is always agreed upon in advance. In the three-hour session it typically is used to help established artists make decisions on what kind of print medium to use, what kind of or which publishers to seek, or what kinds of marketing, advertising and tradeshow strategies would work best for them if they are planning to self-publish their images. The one-hour session normally focuses on getting started.

Rates - A standard rate applies for the one-hour session while a 20% discount is applied to the three-hour session. A contract and retainer for periods longer than three hours will be required. Consulting fees are paid in advance and in full. All major credit cards are accepted.

Details - You incur no phone charges. You will be called at a pre-determined time. It is suggested that the 3-hour Session be done in two 90-minute meetings. If necessary, they can be scheduled in three 60-minute meetings.

A information gathering worksheet that is provided in advance must be completed and returned prior to the first session so that less time is taken getting basics understood, leaving more time for working on your specific needs and questions.

Speaking Engagements

Barney Davey is available for speaking engagements on the subject of art marketing, art print marketing and e-publishing. Sessions from one hour to a full day can be conducted over the Web, by tele-conference and in-person. Please call or email with your questions regarding how you can reserve time on Barney's speaking schedule.

Resources

Trade Publications

Art Business News • Pfingsten Publishing LLC
6000 Lombardo Center Drive, Suite 420, Seven Hills, OH 44131
Phone: 888-772-8926 • 216-328-8926 • Fax: 216-328-9452
www.artbusinessnews.com

Art World News • Wellspring Communications, Inc.
143 Rowayton Avenue, Rowayton, CT 06853 Phone: 203-854-8566
• Fax: 203-854-8569 *www.artworldnews.com*

DECOR • Pfingsten Publishing LLC
1801 Park 270 Dr., Maryland Heights, MO 63146
Phone: 314-824-5500 • FAX: 314-824-5640
www.decormagazine.com

License! • Advanstar Communications, Inc.
One Park Avenue, New York, NY 10016
Phone: 212-951-6600 • Fax: 212-951-6714
www.licensemag.com

Picture Framing Magazine • Hobby Hill Publishing
P.O. Box 102 - 207 Commercial Court, Morganville, NJ 07751
Phone: 732-536-5160 • Toll-Free: 800-969-7176 • Fax: 732-536-5761
www.pictureframingmagazine.com

Volume • Pfingsten Publishing LLC
6000 Lombardo Center Drive, Suite 420, Seven Hills, OH 44131
Phone: 888-772-8926 • 216-328-8926 • Fax: 216-328-9452
www.volumeframingmagazine.com

Tradeshows

Art & Framing Showcase
www.marketplaceexpos.com

Artexpo Atlanta
www.internationalartandframing.com

Artexpo New York
www.internationalartandframing.com

DECOR Expo Atlanta
www.internationalartandframing.com

DECOR Expo New York
www.internationalartandframing.com

Fine Art Forum
www.fine-art-forum.com

Licensing International 2005
www.licensingshow.com

Photo Marketing Association
www.pmai.org

PPFA/PMA Show
www.pmai.org

West Coast Art & Frame Show
www.wcafshow.com

Surtex
www.surtex.com

Consumer Publications

Southwest Art Magazine
5444 Westheimer, Ste. 1440, Houston, TX 77056
713-296-7900 • Fax: 713-850-1314
www.southwestart.com

Wildlife Art Magazine
1428 E. Cliff Rd., Burnsville, MN 55337
Phone: 800-221-6547 • Fax: 952-736-1030
www.wildlifeartmag.com

Art Websites

www.absolutearts.com	Marketplace for art, news, research, gallery and artist portfolios
www.allposters.com	Online Art Retailer
www.art.com	Online Art Retailer
www.barewalls.com	Online Art Retailer
www.fine-art.com	Internet Art Website
www.ebsqart.com	Internet Art Website
www.sedonaartscenter.org	Art Education
www.yessy.com	Internet Art Website

Artist Websites Mentioned

Arnold Friberg
www.fribergfineart.com

Bob Timberlake
www.bobtimberlake.com

Cao Yong
www.caoyongeditions.com

Christian Riese Lassen
www.lassenart.com

Flavia Weedn
www.flavia.com

George Sumner
www.sumner-studios.com

Jane Wooster Scott
www.woosterscott.com

Jody Bergsma
www.bergsma.com

Marty Bell
www.martybell.com

Mary Engelbreit
www.maryengelbreit.com

P. Buckley Moss
www.pbuckleymoss.com

Richard Thompson
www.thompsonsfineart.com
www.impressionistprints.com

Robert Lyn Nelson
www.robertlynnelson.com

Stephen Schutz
www.bluemoutain.com

Thomas Kinkade
www.thomaskinkade.com.

Wyland
www.wyland.com

Yuroz
www.yurozart.com

Government Services

Register art with U.S. Copyright Office
www.copyright.gov/register/visual.html

S.C.O.R.E. (Service Corps of Retired Executives
www.score.org

U.S. Copyright Office
www.copyright.gov

.Information Sources

Decor Sources
www.decor-sources.com

Dictionary.com
www.dictionary.com

International Home Furnishing Center
www.ihfc.com

Wilhelm Research
www.wilhelm-research.com

Marketing

www.ClickXchange.com
Link Exchange

www.GoodKeywords.com
Search Engine Optimization

https://adwords.google.com/select/
Google pay per click marketing

www.LinkExchange.com
Link Exchange

http://www.localsearchguide.org/
Search Engine Optimization

www.ModernPostcard.com
Postcard printing, mailing & lists

www.MySmartMarketing.com
Email and Website marketing

www.passionatepostcarder.com/
Postcard mailing program

http://smallbusiness.yahoo.com/marketing
Yahoo Local Listings

Organizations

Art Copyright Coalition
www.artcc.org

Art Publishers Association
http://apa.pmai.org.

Color Marketing Organization
www.colormarketing.org

Giclée Printers Association
www.gpa.bz

Photo Marketing Association
www.pmai.org

Professional Picture
Framers Assocation
www.ppfa.com

Publicity

Art Deadline	*www.artdeadline.com*
Art Site Guide	*www.artsiteguide.com*
E-Releases	*www.ereleases.com*
Non Starving Artists	*www.nonstarvingartists.com*
PR Newswire	*www.prnewire.com*
PR Web	*www.prweb.com*

Publishers Mentioned

Bentley Publishing Group	*www.bentleypulishinggroup.com*
Grand Image	*www.grandimage.com*
Greenwich Workshop	*www.greenwichworkshop.com*
Haddad's Fine Art	*www.haddadsfinearts.com*
Hadley House	*www.hadleyhouse.com*
Image Conscious	*www.imageconscious.com*
Image Conscious Submission info	*www.imageconscious.com/a/ submissions.html*
Joan Cawley Gallery	*www.jcgltd.com*
Mill Pond Press	*www.millpondpress.com*
Palatino Editions	*www.palatino.net*
Somerset House	*www.somersethouse.com*
Triad Publishing	*www.royoart.com*
Wild Apple Graphics	*www.wildapple.com*
Wild Wings	*www.wildwings.com*

Services

www.2checkout.com	Online Payments
www.artaffairs.com	Website builder
www.artmarketing.com	Mailing lists, books, services for artists
www.artistcareertraining.com	Art Education
www.cj.com	Affiliate Marketing
www.colorqinc.com	Fine art offset lithographic printer
www.ConstantContact.com	Email marketing and management
www.the-dma.org	Privacy Policy Generator
www.franklincovey.com	Mission Statement Builder
www.liebermans.net	Poster consolidator
www.modernpostcard.com	Post card printing mailing & list services
www.mysmartmarketing.com	Website builder & host
www.overture.com	Pay per click marketing
www.paypal.com	Online Payments
www.porterfieldsfineart.com	Licensing Agent & Licensing articles

Software

Business Plan Pro
Palo Alto Business Plan Pro

Working Artist - Artist database software
www.workingartist.com

ACT! – Contact manager software
www.act.com

Filemaker Pro – Easy-to-use database software
www.filemaker.com

Marketing Artist – Free contact/inventory manager
http://marketingartist.com.

Open Source 2.0 - Free Office Suite
www.opensource.org.

Books

Art Licensing 101
Michael Woodward

Art Marketing Handbook
Calvin Goodman

Business and Legal Forms for Fine Artists
Tad Crawford

Chronicles, Vol. 1
Bob Dylan

Electronic Highway Robbery: An Artist's Guide to Copyrights in the Digital Era
Mary E. Carter

Getting Past No: Negotiating Your Way from Confrontation to Cooperation
William Ury

Getting Slightly Famous
Steven Yoder

Guerilla Negotiating
Conrad Levinson, Mark S. A. Smith, Orvel Ray Wilson

How to Profit from the Art Print Market
Barney Davey

In Search of Excellence: Lessons from Americas Best Run Companies
Tom Peters, Bob Waterman

Internet 101 for the Fine Artist with a special guide to Selling Art on eBay
Constance Smith and Susan F. Greaves

Legal Guide for the Visual Artist
Tad Crawford

Licensing Art & Design
Caryn Leland

Man's Search For Meaning
Viktor Frankl

Small Business: An Entrepreneur's Business Plan
J. D. Ryan, Gail Hiduke

The Business of Being An Artist
Daniel Grant

The Definitive Business Plan
Richard Stutely

The Entrepreneur's Success Kit
Kaleil Isaza Tuzman

The Seven Habits of Highly Effective People
Stephen Covey